# Advances and Applications in Electroceramics II

# Advances and Applications in Electroceramics II

## Ceramic Transactions, Volume 235

Edited by
K. M. Nair
Shashank Priya

A John Wiley & Sons, Inc., Publication

Published by John Wiley & Sons, Inc., Hoboken, New Jersey.
Published simultaneously in Canada.

For general information on our other products and services or for technical support, please contact our Customer Care Department within the United States at (800) 762-2974, outside the United States at (317) 572-3993 or fax (317) 572-4002.

Wiley also publishes its books in a variety of electronic formats. Some content that appears in print may not be available in electronic formats. For more information about Wiley products, visit our web site at www.wiley.com.

*Library of Congress Cataloging-in-Publication Data is available.*

ISBN: 978-1-118-27335-7
ISSN: 1042-1122

Printed in the United States of America.

10  9  8  7  6  5  4  3  2  1

# Contents

## MAGNETOELECTRIC MULTIFERROIC THIN FILMS AND MULTILAYERS

## MULTIFUNTIONAL OXIDES

# Preface

New areas of materials technology development and product innovation have been extraordinary during the last few decades. Our understanding of science and technology of the electronic materials played a major role in meeting the social needs by developing innovative devices for automotive, telecommunications, military and medical applications. There is continued growth in this area and electronic technology development has enormous potential for evolving societal needs. Rising demands will lead to development of novel ceramics materials which will further the market for consumer applications. Miniaturization of electronic devices and improved system properties will continue during this decade to satisfy the requirements in the area of medical implant devices, telecommunications and automotive markets. Cost-effective manufacturing should be the new area of interest due to the high growth of market in countries like China and India. Scientific societies should play a major role for development of new manufacturing technology by working together with international counterparts.

One such new materials technology is magnetoelectrics. The realization of a material with simultaneous presence of strong coupling between electric and magnetic order, termed as 'magnetoelectrics' (MEs), would be a milestone for modern electronics and multifunctional materials. It will open the gateway for very high density memory storage media using both magnetic and electric polarization, and the possibility of electrically reading or writing magnetic memory devices (and vice versa). Conversion of magnetic field to electric polarization (direct ME effect or DME effect) or electric field to magnetization (converse ME effect or CME effect) is attractive for various other applications such as magnetic field sensors, tunable phase shifters and filters, and optical components.

The papers in this symposia addressed the developments in synthesis of single phase materials, sintered composites, and bonded composites. Investigations have revealed the presence of both ferroelectricity and magnetism in number of materials such as perovskite type $BiFeO_3$, $BiMnO_3$, the boracite family, $BaMF_4$ compounds (M, divalent transition metal ions), hexagonal $RMnO_3$ (R, rare earths), and the rare earth molybdates, but none seem to provide large coupling between them. Thus, there was constructive discussion on how to address this challenge. The papers included in this proceeding reflect some possible approaches in this direction. The ME effect could be several orders of magnitude larger in piezoelectric – ferromagnetic laminate composites (2-2 connectivity). Especially, the laminate composites consisting of piezoelectric layer and magnetostrictive metal alloy (such as $Tb_{1-x}Dy_xFe_2$ (Terfenol-D) and Fe-Si-B (Metglas)) layer have exhibited strongest ME response over wide frequency range. However, there remains challenge in sintering these different materials together. The presentations in thissymposia addressed this important topic too and showed some sintering approaches. We feel that community has made significant advances in developing both single phase and composite magnetoelectrics.

The Materials societies understand their social responsibility. For the last many years, The American Ceramic Society has organized several international symposia covering many aspects of the advanced electronic material systems by bringing together leading researchers and practitioners of electronics industry, university and national laboratories. Further, The American Ceramic Society has been aggressive in knowledge dissemination by publishing the proceedings of the conferences in the Ceramic Transactions series, a leading up-to-date materials publication, and posting news releases on its website.

This volume contains a collection of 25 papers from three symposia that were held during the 2011 Material Science and Technology Conference (MS&T'11) held at the Columbus Convention Center, Columbus, Ohio USA, October 16–20, 2011. These symposia include: "Advances in Dielectric Materials and Electronic Devices," "Magnetoelectric Multiferroic Thin Films and Multilayers," and "Multifuntional Oxdes". MS&T'11 was jointly sponsored by The American Ceramic Society (ACerS), the Association of Iron & Steel Technology (AIST), the ASM International, the Minerals, Metals & Materials Society (TMS) and the National Association of Corrosion Engineers (NACE).

The editors, acknowledge and appreciate the contributions of the speakers, co-organizers of the three symposia, conference session chairs, manuscript reviewers and Society officials for making this endeavor a successful one.

K. M. Nair, *E.I. duPont de Nemours & Co., Inc., USA*
Shashank Priya, *Virginia Technical Institute & State University, USA*


**SYMPOSIA ORGANIZERS**

**Advances in Dielectric Materials & Electronic Devices**
K.M. Nair, E.I.DuPont de Nemours & Co., Inc
Danilo Suvorov, Jozef Stefan Institute
Ruyan Guo, University Texas at San Antonio
Rick Ubic, Boise State University
Amar S. Bhalla, University of Texas

**Magnetoelectric Multiferroic Thin Films and Multilayers**
Shashank Priya, Virginia Technical Institute & State University
Paul Clem, Sandia National Laboratories
Chonglin Chen, University of Texas at San Antonio
Dwight Viehland, Virgina Technical Institute & State University
Armen Khachaturyan, Rutgers University

**Multifunctional Oxides**
Xiaoqing Pan, University of Michigan
Quanxi Jia, Los Alamos National Laboratory
Pamir Alpay, University of Connecticut
Chonglin Chen, University of Texas at San Antonio
Amit Goyal, Oak Ridge National Laboratory

# Dielectric Materials and Electronic Devices

# DIELECTRIC II-VI AND IV-VI METAL CHALCOGENIDE THIN FILMS IN SILVER COATED HOLLOW GLASS WAVEGUIDES (HGWS) FOR INFRARED SPECTROSCOPY AND LASER DELIVERY

Carlos M. Bledt[1], Daniel V. Kopp[1], and James A. Harrington[1]
[1] Rutgers, the State University of New Jersey
  Department of Materials Science & Engineering
  Piscataway, NJ 08854 USA

ABSTRACT

Hollow glass waveguides (HGWs) are an attractive alternative to traditional photonic bandgap fibers and other infrared fibers for use in various applications involving optimal transmission in the Long Wavelength Infrared (LWIR) region due to their inherent broadband transmission and easily customizable properties. The use of II-VI and IV-VI binary metal chalcogenide thin films in silver coated silica Hollow Glass Waveguides deposited via dynamic liquid phase deposition (DLPD) for infrared spectroscopy and laser delivery has allowed for maximal signal throughput with high laser power thresholds while simultaneously retaining superior single-mode like $TEM_{00}$ properties. The methodology for the aqueous chemical deposition of metal chalcogenide thin films in silver coated silica hollow waveguides including cadmium sulfide (CdS) and lead sulfide (PbS) is presented in this study along with their optical response as determined primarily through Fourier Transform Infrared (FTIR) spectroscopy and infrared ($CO_2$) laser analysis.

## INTRODUCTION

Dielectric coated hollow glass waveguides (HGWs) are capable of low-loss, broadband delivery of electromagnetic radiation at infrared wavelengths typically ranging from $1.0 - 15.0$ μm. In comparison to other types of available infrared fibers HGWs have the advantages of having no end reflections, high laser power throughput threshold, high chemical and mechanical stability, and low beam divergence.[1] Furthermore, HGWs can be engineered for maximum broadband transmission at desired wavelength ranges through simple control of fabrication methodology variables. The physical structure of HGWs consists of a high-purity silica capillary tube of fixed diameter with a protective polyimide or UV-acrylate outer coating. The optically functional structure is created through the deposition of a reflective film and subsequent dielectric thin film(s) on the inner surface of the silica capillary deposited from component containing aqueous solutions. HGW bore sizes can furthermore be chosen with the desired application(s) in mind, such as the necessity for low order mode propagation or high radiation intensity throughput. Figure 1 depicts a representative general diagram of an HGW.

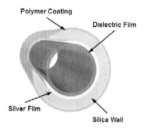

Figure 1. Representative cross-sectional diagram of an HGW

Losses in HGWs depend on a number of factors including bore size, applied bending radius, thin film materials used, coupling efficiency, and propagating modes.[1]

The key in developing low-loss HGWs lies in the manipulation of the fabrication methodology so as to deposit high-quality dielectric films with proper thickness for maximum transmission at desired wavelength ranges as necessary. Traditionally, HGWs have incorporated silver iodide (AgI) as a dielectric film due to the ability to fabricate high-quality AgI thin films in HGWs through mass transport driven subtractive iodization of the pre-deposited silver film.[1] Alternatively, research in the development of HGWs has focused on the use of other IR transparent dielectric materials as functional thin films. Of such possible materials, the II-VI and IV-VI metal chalcogenides such as cadmium sulfide (CdS) and lead sulfide (PbS) are of particular interest due to several factors, including their high optical transparency at IR wavelengths, the vast range of refractive indices covered by these materials, and their ability to be deposited via similar electroless chemical deposition methods. Of particular interest is the use of chalcogenide thin films in multilayer dielectric HGW designs, in which metal chalcogenide films of materials with alternating low and high refractive indices are deposited to form a periodic alternating dielectric constant resulting in a low-loss, omnidirectional reflecting 1-D photonic bandgap structure.[2]

EXPERIMENTAL METHODOLOGY

The successful fabrication of high-quality metal chalcogenide thin films via electroless deposition from aqueous solutions relies on the simultaneous controlled release of reactive metallic ion species and free chalcogenide ions in solution. Metal chalcogenide thin films are deposited in HGWs through such electroless deposition methods via dynamic liquid phase deposition (DLPD), which is an adaption of chemical bath deposition (CBD) where the reactive solution is continuously pumped by a peristaltic pump through the pre-silver coated HGW silica capillary tube. As such, volumetric fluid flow rate poses an additional processing variable in DLPD which along with the main governing variables of component concentrations, complexing agents used, temperature, and solution pH found in CBD deposition of metal chalcogenide thin films directly influence the quality and growth rate of the films. Figure 2 shows the general DLPD processing setup involving the simultaneous pumping of unreacted constituent component solutions at equal rates through the pre-silver coated HGW.

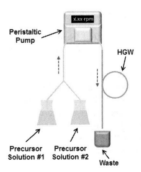

Figure 2. General DLPD procedure setup for fabricating HGWs

In comparison to CBD methods, DLPD has the advantage that substantially thicker films can be deposited due to the fact that the continuous flow of unreacted solutions prevents the precursor depletion and film saturation witnessed in CBD film growth methods.

Deposition of Reflective Silver Thin Films

Initial fabrication of HGWs involves the deposition of a highly reflective silver film on the inner surface of the silica capillary via DLPD methodology. The silver deposition procedure is preceded by a short sensitization step, in which the inner silica surface is 'activated' by pumping an acidic 1.60 mM tin (II) chloride solution. The sensitization procedure allows the subsequent silver film deposition procedure time to be considerably shortened due to the initial rapid reduction of complex silver ions by surface adhered tin (II) ions which effectively reduces the initiation time for the formation of the silver film. This in turn allows the total silver film deposition time to be shortened, thus reducing surface roughness seen with prolonged silver film deposition times which has been shown to increase scattering losses.[3] The silver thin film deposition procedure involves the simultaneous flow of an alkaline ammonia complexed 14.4 mM silver (I) nitrate at a pH of 11.0 – 11.4 solution and a 3.11 mM dextrose reducing solution. Reduction of the soluble silver (I) diammine cation by the aldehyde functional group containing dextrose in solution results in the deposition of the metallic silver film on the inner capillary silica surface. The deposited silver film must be thicker than the penetration depth of the wavelength for which the HGW will be used while at the same time being sufficiently thick to function as a suitable substrate for the deposition of subsequent metal chalcogenide dielectric thin films. In practice the thickness of the deposited silver film varies from 100 – 300 nm as necessary, corresponding to silver deposition times ranging from 2 – 15 minutes. All deposition procedures were carried out utilizing a standardized peristaltic pump speed which gave a steady volumetric flow rate of 17.35 mL/min and carried out under ambient conditions at 25 °C.

Deposition of Dielectric Cadmium Sulfide Thin Films

The successful deposition of high quality CdS thin films in HGWs requires careful control of the fabrication methodology variables, particularly component concentrations, solution pH, and volumetric fluid flow rate. The CdS deposition procedure involves simultaneous flow of an alkaline ammonia complexed cadmium ion solution and a thiourea solution as a sulfide ion source. The reaction mechanism involved in the CdS deposition procedure involves the successful complexing of a cadmium (II) nitrate solution through addition of excess ammonium hydroxide, thus creating a stable ammonia complexed cadmium ion solution to allow for controlled availability of cadmium ion species in solution. Simultaneous controlled release of sulfide ions in solution is likewise essential, with the controlled hydrolysis of dissolved thiourea in alkaline medium determining the availability of reactive sulfide ions in solution. The reaction mechanisms involved in the deposition of CdS from alkaline ammonia complexed ion and thiourea containing solutions has been discussed by several authors in the literature.[4,5,6] Primarily, two competing simultaneous deposition mechanisms have been proposed with the corresponding chemical reactions involved in the formation of CdS being presented in Equations 1 – 4.[4,6]

$$SC(NH_2)_2 + 2OH^- \rightleftharpoons S^{2-} + CN_2H_2 + 2H_2O \tag{1}$$

$$Cd^{2+} + 4NH_3 \rightleftharpoons Cd(NH_3)_4^{2+} \tag{2}$$

$$Cd(NH_3)_4^{2+} + 2OH^- + site \rightarrow [Cd(OH)_2]_{ads} + 4NH_3 \tag{3a}$$

$$[Cd(OH)_2]_{ads} + S^{2-} \rightarrow CdS_{(s)} + 2OH^- \tag{3b}$$

$$Cd(NH_3)_4{}^{2+} + S^{2-} \rightarrow CdS_{(s)} + 4NH_3 \tag{4}$$

The two competing mechanisms involved in the deposition of CdS films shown by Equation 3 and Equation 4 show the dilemma involved between depositing high quality CdS films of considerable thicknesses and reducing the prolonged deposition times involved in the fabrication of these films. Equation 3 shows the proposed heterogeneous ion by ion CdS film deposition mechanism in which intermediate cadmium (II) hydroxide is adsorbed on the substrate surface and is subsequently transformed to well adhering high quality CdS thin films.[4,6,7] Equation 4 shows the proposed competing homogeneous cluster by cluster CdS film deposition mechanism in which colloidal CdS particles agglomerate on the substrate to form low quality, high surface roughness, porous CdS films. While the heterogeneous deposition mechanism is preferred for the deposition of high quality CdS thin films, this reaction mechanism is much slower and can be inconvenient as the sole CdS film deposition mechanism for CdS thin films of thicknesses greater than 300 nm. The homogeneous deposition mechanism suggested by Equation 4 results in much faster film growth rates but at the expense of decreased film uniformity and quality.[5] The necessity of depositing high quality CdS films in HGWs of thicknesses greater than 300 nm while at the same time limiting the deposition procedure time suggests a careful balance between the competing heterogeneous and homogeneous deposition mechanisms must be found. Optimization of the CdS film deposition procedure involved utilizing equivalent parts of a 14.98 mM cadmium (II) nitrate, 2.0 M ammonium hydroxide solution and a 150 mM thiourea solution at a pH of 11.45 – 11.65. The experimental determination of the CdS film growth kinetics in HGWs involved fabrication of several samples coated at different deposition times ranging from 150 to 360 minutes in fixed 30 minute intervals.

Deposition of Dielectric Lead Sulfide Thin Films

The successful deposition of high quality PbS thin films in HGWs likewise requires careful control of several fabrication methodology variables such as component concentrations, solution pH, and volumetric fluid flow rate. Thiourea was again used as a sulfide ion source, with the complexing agent of choice for the free lead (II) ions being hydroxide ions introduced as a sodium hydroxide solution in place of ammonium hydroxide in the case of CdS deposition. The addition of excess sodium hydroxide to a solution of lead (II) cations results in the formation of plumbite complex ions. Complexing the lead (II) ions in this manner allows for controlled release of available lead (II) reactive species in solution, preventing spontaneous settling of PbS particles from solution and allowing for the controlled deposition of high-quality PbS thin films. The proposed PbS thin film deposition mechanism is presented in Equations 5 – 7.[4,6]

$$SC(NH_2)_2 + 2OH^- \rightleftarrows S^{2-} + CN_2H_2 + 2H_2O \tag{5}$$

$$Pb^{2+} + 6OH^- \rightleftarrows Pb(OH)_6{}^{2-} \tag{6}$$

$$Pb(OH)_6{}^{2-} + site \rightarrow [HPbO_2{}^-]_{ads} + 3OH^- + H_2O \tag{7a}$$

$$[HPbO_2{}^-]_{ads} + S^{2-} \rightarrow PbS_{(s)} + H_2O \tag{7b}$$

It is important to note that in the case of PbS thin films, heterogeneous deposition mechanisms have been widely regarded as being predominant over homogeneous deposition mechanisms at high pH values.[4] This fact allows for the rapid deposition of high quality PbS films at substantially higher thicknesses than those currently achievable for CdS films. Optimization of the PbS film deposition procedure involved utilizing equivalent parts of a 5.43 mM lead (II) nitrate, 56.3 mM sodium hydroxide solution and a 54.4 mM thiourea solution at a pH of 12.05 – 12.15. The experimental determination of the PbS film growth kinetics in HGWs involved fabrication of several samples coated at different deposition times ranging from 15 to 135 minutes in fixed 15 minute intervals.

RESULTS

FTIR Spectroscopic Analysis

Initial analysis of the fabricated CdS and PbS coated HGW samples involved FTIR spectroscopic analysis of each of the individual segments processed for the various deposition times. FTIR analysis was carried out using a Bruker Tensor 37 FTIR spectrometer in conjunction with a Teledyne Judson cryogenic MCT/A detector. The FTIR beam path was diverted and focused into the input end of the 15 cm long coated 700 μm ID HGW sample using a right angle silver coated parabolic mirror. The output FTIR signal from the HGW sample segment was collected using the MCT/A detector, thus making possible the qualitative and quantitative analysis of the spectral response of the various fabricated metal chalcogenide coated samples. The spectral response for each of the 16 samples was measured in this manner and the IR spectral response from λ = 2.0 – 15.0 μm for select samples is shown in Figure 3 for both metal chalcogenide deposition procedures.

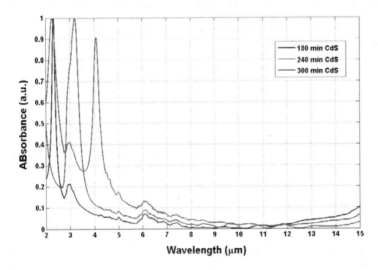

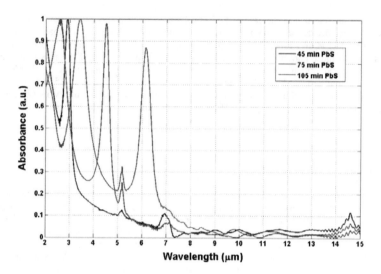

Figure 3. Spectra of a) CdS (180, 240, 300 min) and b) PbS (45, 75, 105 min) coated HGW samples

The FTIR spectra for all samples show the definitive thin film interference peaks due to the deposited chalcogenide thin films. As expected, the interference peaks shift towards longer wavelengths with increasing deposition times due to increasing film thicknesses. For the deposition times studied, all deposited films PbS films were of high-quality as depicted by the well-defined interference peaks and narrow nature of the first interference peak. The CdS films were likewise seen to be of high optical quality for most of the studied deposition times yet considerable non-uniformity was seen in film deposition times longer than 360 minutes. It is proposed that the non-uniformity of thicker CdS films may be due to increasing surface roughness with time as a result of prolonged homogenous thin film formation.[8]

Thin Film Deposition Kinetics

The thicknesses of the deposited films can be calculated from the spectral data by taking into account the centroid wavelength of the $1^{st}$ interference peak ($\lambda_p$) as well as the refractive index, n, of the dielectric material using Miyagi's formula (Equation 8) based on thin film optics theory.[2]

$$d_{film} = \frac{\lambda_p}{4\sqrt{n^2 - 1}} \qquad (8)$$

The film thickness for the CdS (n=2.27)[9] and PbS (n=4.00)[9] deposition procedures as a function of time shown in Figure 4 and was determined by calculating the film thicknesses at each of the corresponding deposition times from the centroid wavelength of the first interference peak for each of the obtained spectral responses.

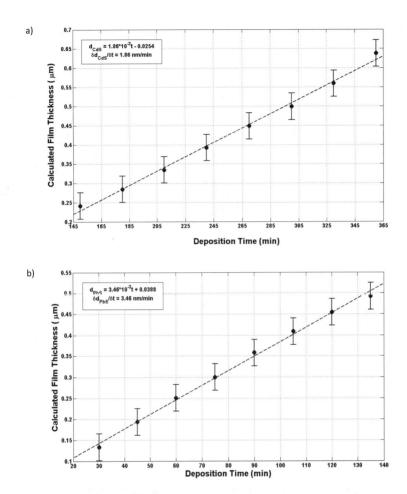

Figure 4. Growth kinetics for DLPD deposited a) CdS and b) PbS films in HGWs

As expected, both the CdS and PbS deposition kinetics data show a strongly linear film growth rate in this region past the non-linear film nucleation region. Utilizing the aforementioned CdS and PbS solution precursor concentrations and fixed pH for each of the solutions, film growths of 1.86 nm/min and 3.46 nm/min were obtained for CdS and PbS films respectively. Furthermore, the deposition times involved in depositing thin films of such thicknesses are acceptable and film thickness control is high. As previously mentioned this is due to the fact that no depletion of precursor species occurs in DLPD as it does in CBD film deposition as unreacted solution is continuously pumped through the specimens. This processing methodology not only allows for a continuously linear film growth rate, but also allows for much more thicker films to be deposited due to the non-existence of precursor depletion and saturation which limit the film thicknesses achievable by CBD methods.

Infrared Transmission Properties

Final analysis of CdS and PbS coated HGWs involved the fabrication of CdS and PbS coated 1.5 m long HGWs with targeted dielectric film thicknesses of approximately 500 nm and 350 nm for CdS and PbS thin films respectively. These samples were fabricated using the methodology outlined in the experimental procedure section with deposition times of 300 minutes (CdS) and 95 minutes (PbS) as determined necessary to obtained these film thicknesses from the growth rate kinetics analysis. These specific film thicknesses were chosen due to the fact that thicker dielectric films are necessary for maximum transmission at longer wavelengths as well as for the fact that good uniformity was maintained at these deposition times as determined from the preceding spectral analysis. The spectral optical response of these samples was taken with the FTIR spectrometer to determine the quality of the deposited CdS and PbS films and is shown in Figure 5.

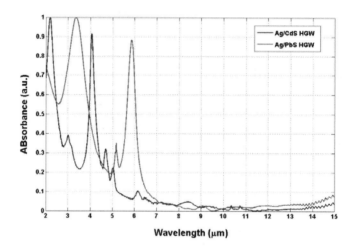

Figure 5. FTIR spectra of Ag/CdS HGW (t = 300 min) and Ag/PbS HGW (t = 95 min)

The spectra of the fabricated samples again shows high quality CdS and PbS films with calculated thicknesses of 497 nm and 378 nm respectively.

The transmissive properties of the fabricated samples were analyzed using a Synrad I-series $CO_2$ 15 Watt max. laser emitting at a wavelength of $\lambda = 10.6$ μm. The samples were properly aligned in the beam path for optimal coupling efficiency using a zinc selenide lens with a focal length of 7.5 inches. The transmitted power for each sample was measured under no applied bending as well as applied bending radii of 2.5, 2.0, 1.5, 1.0, 0.75, and 0.5 m using an Ophir Vega power meter. The length of sample under bending was kept constant at 80 cm for all bending measurements. The losses in dB/m were calculated using a cutback methodology to reduce the possibility of experimental error due to coupling inefficiencies. The mathematical equation for determining the loss is a derivation of the Beer-Lambert law and is given in Equation 9.

$$\alpha \,[dB/m] = \frac{10}{L-s} \log_{10}\left(\frac{P_{in}}{P_{out}}\right) \tag{9}$$

where L is the length in meters of the entire sample, s is the length of the cutback segment, $P_{in}$ is the measured power out of the cutback segment, and $P_{out}$ is the measured power out of the entire sample length. Utilizing this methodology, the loss of the samples as a function of curvature was determined and presented in Figure 6.

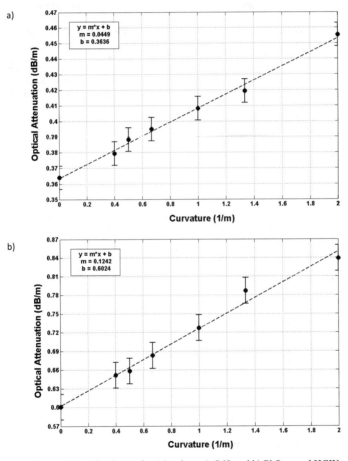

Figure 6. Bending losses for 1.5 m long a) CdS and b) PbS coated HGWs

The losses were overall higher for the Ag/PbS waveguide than for the Ag/ CdS waveguide as would be expected from theory due to the higher refractive index of PbS relative to CdS. Furthermore, the loss increase upon applied bending shows the expected linear 1/R dependency and it is evident the bending loss increase of the Ag/PbS waveguide is higher than that for the Ag/CdS waveguide at $m_{PbS} = 0.124$ vs. $m_{CdS} = 0.045$ respectively. Overall, both metal chalcogenide films were proven to be successful in substantially reducing the loss in comparison to Ag only coated HGWs with typical straight losses of approximately 3.0 dB/m.

CONCLUSION

The successful deposition of high-quality cadmium sulfide and lead sulfide in HGWs via DLPD deposition methodology has been demonstrated in this study. The reaction mechanisms involved in the deposition of these metal chalcogenide thin films have been reviewed in order to gain an understanding of the processes involved to allow for the deposition of high-quality optical thin films and achieve the necessary film thickness deposition control needed to design both single layer as well as multilayer dielectric coated low-loss HGWs. The film growth kinetics at the deposition parameters presented show relatively rapid linear film growth in comparison to traditional CBD CdS and PbS methods. Spectral analysis resulting from the film deposition kinetics study shows optical grade CdS thin films up to thicknesses of approximately 600 nm and optical grade PbS thin films upwards of thicknesses of 600 nm via the aforementioned processing methodology. Infrared attenuation measurements carried out on the fabricated samples show both metal chalcogenide films are effective at substantially reducing losses in HGWs when compared to Ag only coated waveguides. Furthermore, the losses in Ag/CdS HGWs were shown to be less than those in Ag/PbS HGWs with a higher loss increase upon applied bending in Ag/PbS HGWs than in Ag/CdS HGWs.

REFERENCES

[1] Harrington, J. A., *Infrared Fiber Optics and Their Applications*, (SPIE Press, Bellingham, WA, 2004).
[2] Miyagi, M. and Kawakami, S. "Design theory of dielectric-coated circular metallic waveguides for infrared transmission," IEEE Journal of Lightwave Technology. LT-**2**, 116-126 (1984).
[3] Rabii, C. D., Gibson, D. J., and Harrington, J. A., "Processing and characterization of silver films used to fabricate hollow glass waveguides," Appl. Opt. **38**, 4486-4493 (1999).
[4] Chaparro, A. M., "Thermodynamic analysis of the deposition of zinc oxide and chalcogenides from aqueous solutions," Chem. Mater., **17** (16), 4118-4124 (2005)
[5] Guillen, C., Martinez, M. A, Herrero, J., "Accurate control of thin film CdS growth process by adjusting the chemical bath deposition parameters," Thin Solid Films, **335**, 37 – 42 (1998).
[6] Niesen, T. P., De Guire, M. R., "Review: Deposition of Ceramic Thin Films at Low Temperatures from Aqueous Solutions." Journal of Electroceramics, **6**, 169 – 207 (201).
[7] R. S Mane and C. D Lokhande, "Chemical deposition method for metal chalcogenide thin films," Mat. Chem. Phys. **65**, 1-31 (2000).
[8] Gopal, V., Harrington, J. A., "Deposition and characterization of metal sulfide dielectric coatings for hollow glass waveguides," Optics Express, **11**, 24 (2003).
[9] Palik, E. D. and Ghosh, G., Handbook of optical constants of solids, (Academic, London, 1998).

* cmbledt@eden.rutgers.edu; Phone 862-485-9289; irfibers.rutgers.edu

DIELECTRIC PROPERTIES OF CHEMICALLY BONDED PHOSPHATE CERAMICS
FABRICATED WITH WOLLASTONITE POWDERS

H. A. Colorado[a,b], A. Wong[a], J. M. Yang[a]
[a]Materials Science and Engineering Department, University of California, Los Angeles, CA 90095, USA
[b]Universidad de Antioquia, Mechanical Engineering. Medellin-Colombia

ABSTRACT
This paper reports on the recent progress towards the development of a dielectric chemically bonded phosphate ceramic (CBPC) matrix composite with $BaTiO_3$ contents. These CBPCs are fabricated at room temperature, starting as a liquid with a controllable setting time, which can make a big impact in the manufacturing of electronic materials.

The CBPCs were made by mixing phosphoric acid formulation and Wollastonite powder using the Thinky Mixing apparatus. Four different Wollastonite sizes were utilized in the manufacturing of CBPCs. Also, barium titanate was added in different concentrations to the samples to see its effects on the dielectric properties.

The characterization was conducted with Scanning Electron Microscopy, X-Ray Diffraction, compressive strength, density and porosity measurements. Results showed this material as a promissory solution for capacitor and other dielectric applications. Also, it was found that the compressive strength increased as $BaTiO_3$ content increased, and that for all samples with $BaTiO_3$, the mean compressive strength was over 100MPa. Density and porosity results, however, showed that as $BaTiO_3$ content increased, the porosity decreased.

INTRODUCTION
Chemically Bonded Ceramics (CBCs) are inorganic solids synthesized by chemical reactions at low temperatures without the use of thermally activated solid-state diffusion (typically less than 300°C)[1,2,3]. This method avoids high temperature processing (by thermal diffusion or melting) which is the normal in traditional ceramics processing. The chemical bonding in CBCs allows them to be inexpensive in high volume production. Because of this, CBCs have been used for multiple applications. They include: dental materials[4], nuclear waste solidification and encapsulation[5,6], and composites with fillers and reinforcements[7,8]. The fabrication of conventional cements and ceramics is energy intensive as it involves high temperature processes and emission of greenhouse gases, which adversely affect the environment[9].

CBCs have been poorly explored as electronic materials, although previous research has shown their potentials, for instance, as insulators with low dielectric constant[10]. It has been shown in cements and ceramics that the water contents change the dielectric properties, which have been used to study, for instance, the Portland cement hydration[11]. Water molecules in the cement paste changed from free to bound water in various states of hydration or crystallization, affecting the bonding state, which was reflected in the dielectric constant.

In this research, Wollastonite powder ($CaSiO_3$) was mixed with a phosphoric acid formulation ($H_3PO_4$) in a ratio of 1.2 $CaSiO_3$ to acidic formulation. Then, $BaTiO_3$ powders were added to the mixture. These products reacted into a Chemically Bonded Phosphate Ceramic (CBPC). The mixing of Wollastonite with the acid formulation produced a composite material with crystalline brushite ($CaHPO_4 \cdot 2H_2O$) and Wollastonite phases; and amorphous silica and calcium phosphates phases[8]. As in Portland cement, the water also played an important role in the binding phases for our CBPC, although the reactions were completely different (acid-base reactions). Water started off in a free state and later became bonded in the produced calcium phosphates.

The sections below will present the Wollastonite based CBPCs with BaTiO$_3$ contents characterized by compression, density, Scanning Electron Microscopy (SEM) and X-Ray Diffraction (XRD).

This research opens up other possibilities to produce a new emerging class of electronic materials, fabricated at room temperature. A multiphase CBPC material could potentially be grown at room temperature having phases with such different properties that would contribute to the whole composite.

EXPERIMENTAL

Samples manufacturing

CBPC samples were fabricated by mixing an aqueous phosphoric acid formulation and natural Wollastonite powder M200 (from Minera Nyco; see Table I) in a 1.2 ratio liquid to powder. Also, BaTiO$_3$ with purity 99% (from Strem Chemicals) was added to the mixture in different concentrations. For all of the samples, the 1.2 ratio liquid (phosphoric acid formulation with pH 1.0) to powders (Wollastonite + BaTiO$_3$) remained constant, see Table II. In all cases, mixed components started by mixing Wollastonite and Phosphoric acid formulation. Then, for samples with BaTiO$_3$, 1 minute of additional mixing was performed. Hence, dielectric properties could be preserved by adding BaTiO3 particles at the end, as their interaction with acidic formation would be limited.

Table I. Chemical composition of Wollastonite powder.

| Composition | CaO | SiO$_2$ | Fe$_2$O$_3$ | Al$_2$O$_3$ | MnO | MgO | TiO$_2$ | K$_2$O |
|---|---|---|---|---|---|---|---|---|
| Percentage | 46.25 | 52.00 | 0.25 | 0.40 | 0.025 | 0.50 | 0.025 | 0.15 |

Table II. Raw materials contents in the composites samples fabricated.

| Raw materials | CBPC | | CBPC-1wt% BaTiO$_3$ | | CBPC-5wt% BaTiO$_3$ | | CBPC-10wt% BaTiO$_3$ | |
|---|---|---|---|---|---|---|---|---|
| | Weight (g) | wt% | Weight (g) | wt% | Weight (g) | wt% | Weight (g) | wt% |
| Wollastonite | 50 | 45.45 | 50 | 45.00 | 50 | 43.29 | 50 | 41.32 |
| Phosphoric acid formulation | 60 | 54.55 | 60 | 54.01 | 60 | 51.95 | 60 | 49.59 |
| BaTiO$_3$ | 0 | 0.00 | 1.1 | 0.99 | 5.5 | 4.76 | 11 | 9.09 |
| TOTAL | 110 | 100 | 111.1 | 100 | 115.5 | 100 | 121 | 100 |

The mixing process of the components was conducted in a Planetary Centrifugal Mixer (Thinky Mixer® AR-250, TM). It has been found that when the temperature of the raw materials was reduced, the pot life of the composite increased and the viscosity decreased, which affected the quality of the composite[12]. Thus, the raw materials for compression samples were first stored in a refrigerator for an hour at 3°C in independent and closed containers. It has been reported[13] that the pot life can be extended through different methods: thermal processing, aging the raw materials, or putting additives during the mixing process.

Dielectric tests

Samples for dielectric tests were fabricated using glass molds of 11.4 mm diameter and 10cm long, which were cut into disks using a diamond saw. Samples were then ground using silicon carbide papers of grit ANSI 400 until they were 1.5mm thick by using a metallic mold. Finally, samples were dried in a furnace at 50°C for 24 hours, and then at 100°C for 24 hours in order to stabilize the water content. An LCR Meter Agilent 16451B dielectric test apparatus was used in this test. The study frequency range was between 20Hz and 1MHz, and all tests were conducted at room temperature.

Compression tests

Samples were fabricated using glass molds of 12.7mm diameter and 100mm long. Then, a diamond saw was used to cut cylinders of 28mm in length. Samples were ground (with silicon carbide papers of grit ANSI 400) using a metallic mold until flat, parallel and smooth surfaces were obtained. The final length was 25.4mm. Samples were dried in a furnace at 50°C for 24h and then at 100°C for 24 hours in order to stabilize the water content. Compression tests were then conducted in an Instron machine 3382. A set of 5 samples was tested for each composition of Table II. The crosshead speed was 1mm/min. Figure 1 shows the compression samples.

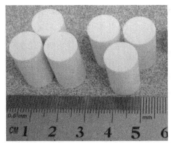

Figure 1. Typical compression samples before being tested.

Other Characterizations

To see the microstructure, sample sections were initially ground using silicon carbide papers grit ANSI 240, 400 and 1200 incrementally, and then they were polished with alumina powders of 1, 0.3 and 0.05μm grain size incrementally. After polishing, samples were dried in a furnace at 50°C for 24h and then at 100°C for 24 hours, and were observed through an optical microscope. For SEM examination, samples were mounted on an aluminum stub and sputtered in a Hummer 6.2 system (15mA AC for 30 sec), creating approximately a 1nm thick film of Au. The Scanning Electron Microscope used was a JEOL JSM-6700F FE-SEM in high vacuum mode. Elemental distribution x-ray maps were collected on the SEM with an energy-dispersive X-ray spectroscopy analyzer (EDAX, aka EDS). The images were collected on the polished and gold-coated samples, with a counting time of 51.2 ms/pixel.

X-Ray Diffraction (XRD) experiments were conducted using an X'Pert PRO (Cu Kα radiation, λ=1.5406 Å), at 45KV and scanning between 10° and 80°. M200, M400 and M1250 Wollastonite samples (before and after the drying process) were ground in an alumina mortar, and XRD tests were done at room temperature.

Density tests were conducted over CBPCs with Fly ash as the filler. All samples were tested after the drying process described above, in a Metter Toledo balance, by means of the buoyancy method. The Dry Weight ($W_d$), Submerged Weight ($W_s$), and Saturated Weight ($W_{ss}$) were measured. Then the following parameters were calculated:

$$\text{Bulk volume: } V_b = (W_{ss} - W_s)/ (1.0 \text{ g/cm}^3) \tag{1}$$

$$\text{Apparent volume: } V_{app} = (W_d - W_s)/ (1.0 \text{ g/cm}^3) \tag{2}$$

$$\text{Open-pore volume: } V_{op} = (W_{ss} - W_d)/ (1.0 \text{ g/cm}^3) \tag{3}$$

$$\text{\% porosity} = (V_{op}/V_b) \times 100 \text{ \%} \tag{4}$$

$$\text{Bulk Density: } D_b = W_d/ (W_{ss} - W_s) \tag{5}$$

$$\text{Apparent Density: } D_a = W_d/(W_d - W_s) \tag{6}$$

In these calculations, density of water was taken to be $1.0 \text{ g/cm}^3$. A set of three samples per composition was tested.

RESULTS
Figure 2a and b shows the morphology of the powders used, Wollastonite and BaTiO$_3$ respectively. Figure 2c and d shows cross section view images for the CBPC and the CBPC with 10wt% BaTiO$_3$. Figure 2c shows the CBPC as a composite material with an amorphous phosphate matrix with grains of silica (SiO$_2$), remaining Wollastonite (CaSiO$_3$) and brushite (CaHPO$_4$·2H$_2$O). Figure 2d shows the CBPC with 5wt% of BaTiO$_3$.

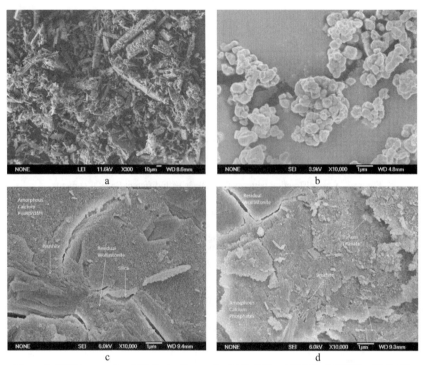

Figure 2. SEM images for a) Wollastonite powder, b) BaTiO$_3$, c) CBPC, d) CBPC with 5wt% BaTiO$_3$.

Figure 3 shows XRD for CBPCs with different BaTiO$_3$ contents. For reference, pure BaTiO$_3$ and the CBPC without BaTiO$_3$ have been included. As BaTiO$_3$ increased in the CBPC, in general, the corresponding intensity of the peaks increased as expected.

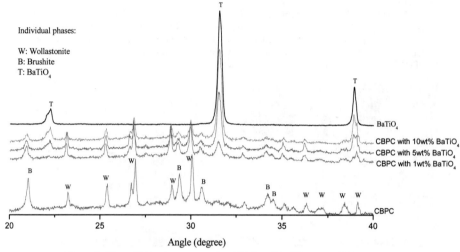

Figure 3. X ray diffraction patterns for CBPCs with different BaTiO₃ contents

Figure 4 shows the compressive strength for the ceramic samples fabricated chemically. It was found that as BaTiO₃ increased, the compressive strength increased. Also, it can be seen that the standard deviation was low. The compressive strength in the CBPC samples with 10wt% of BaTiO₃ is about three times the strength of the CBPC samples, which is significant for applications where components are under high stresses.

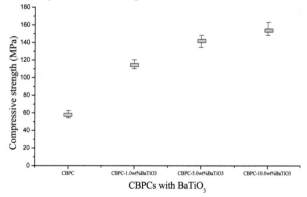

Figure 4. Compressive strength results for CBPCs with different BaTiO₃ contents.

Figure 5 shows typical curves, as part of the results presented in Figure 4. It can be seen that when BaTiO₃ was added, the compressive strength increased, and went over 100MPa in all cases.

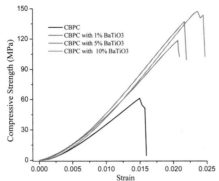

Figure 5. Typical compression stress-strain curves for the ceramic composites fabricated.

Table III shows density and porosity results for the samples fabricated. It can be observed that as BaTiO₃ content increased the porosity decreased.

Table III. Density and porosity results for the CBPC with different BaTiO₃ contents

| Sample composition | | $D_b$ [g/cm³] | $D_a$ [g/cm³] | % Poros. |
|---|---|---|---|---|
| CBPC | Mean values | 1.56 | 1.83 | 14.8 |
| | Stand. Dev. | 0.01 | 0.01 | 0.0 |
| CBPC with 1wt% BaTiO₃ | Mean values | 1.94 | 2.27 | 14.5 |
| | Stand. Dev. | 0.04 | 0.02 | 0.7 |
| CBPC with 5wt% BaTiO₃ | Mean values | 1.95 | 2.26 | 13.9 |
| | Stand. Dev. | 0.01 | 0.00 | 0.4 |
| CBPC with 10wt% BaTiO₃ | Mean values | 2.02 | 2.25 | 10.2 |
| | Stand. Dev. | 0.05 | 0.02 | 1.6 |

Figure 6 shows the relative dielectric constant as a function of the frequency for the CBPCs fabricated.

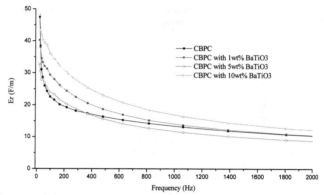

Figure 6 Relative dielectric constant for CBPCs with different $BaTiO_3$ contents

Figure 7 shows the dissipation factor (tan δ) as a function of the frequency for the CBPCs fabricated.

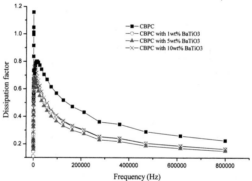

Figure 7 Dissipation factor (tan δ) for CBPCs with different $BaTiO_3$ contents

DISCUSSION

As shown in Figure 2C, the CBPC's SEM image shows features of an amorphous matrix with grains of silica ($SiO_2$), remaining Wollastonite ($CaSiO_3$),and brushite ($CaHPO_4 \cdot 2H_2O$). From the XRD graph shown in Figure 3, the composite samples displayed a mixture of the corresponding peaks from both the $BaTiO_3$ and the CBPC. Although there were no new additional peaks shown in the compounds, it is possible that amorphous phases were created instead of crystalline sub-products and as a result, would not show up as a peak. For example if the $BaTiO_3$ reacted, it could be in amorphous phases, such as barium or titanium phosphate. This reaction will happen when either the barium, titanium, or both diffuse out by the phosphate action and act as a binding phase. Thus, this part of the research should be further investigated to determine compositional breakdowns of the final product.

Regarding the compressive strength, there was progressively an increase in the strength as more $BaTiO_3$ was mixed with CBPC as shown in Figure 4. For example in the CBPC samples with 10 wt% of $BaTiO_3$, the compressive strength improved by three folds compared to the reference sample, which is significant. In this case, the $BaTiO_3$ particle was acting as a reinforcement and possibly new phases were being formed (i.e. barium or titanium phosphates) that improved the overall compressive strength. In addition to the increase in compressive strength, there was an overall increase in density when comparing the composites the reference sample as well as a decrease in porosity as shown in Table III. These trends suggest that the $BaTiO_3$ is filling voids.

For all compositions, the dielectric constant remained fairly constant in the study frequency range (20Hz-1MHz), although only the low region was considered as shown in Figure 6. Only for CBPC samples with 10 wt% of $BaTiO_3$ a considerable increase respect to other samples is observed. For other samples, no further increase was observed as concentrations of $BaTiO_3$ were increased. It is important to remember from the experimental procedure, that all samples were dried in a furnace at 50°C for 24 hours and then at 100°C for 24 hours in order to stabilize the water content. Results for non-thermally treated samples, as well as other processing approaches (functionalized $BaTiO_3$ particles, micro fibers and growing new phases), over these materials are currently under improvement and will be presented in a future paper.

The potential of the CBPCs for the electronic materials can be highlighted in two main points. The first point is that CBPCs are versatile composite materials that can design multiphase ceramic materials with selected properties assigned to specific phases. However, it should be noted that this variability itself can represent a problem for large scale manufacturing. The second one is that these CBPCs can be fabricated at room temperature where they start as a liquid with a controllable setting time, which provides another advantage from a manufacturing perspective. By processing the material as a liquid at room temperature, it is easier to make any changes or additions to tailor the electronic properties of the sample.

Thus, this research opens up other possibilities to produce a new emerging class of electronic ceramics materials fabricated at room temperature. This low temperature process can be applied not only to phosphate ceramics, but also other ceramics like $BaTiO_3$ or other ceramics in electronics that were traditionally fabricated at high temperatures. One potential field could be multiferroics, such as materials with simultaneous coexistence of ferroelectric, ferromagnetic and ferroelastic phases[14]. A multiphase material could potentially be grown via CBPC manufacturing, having phases with such different properties that would contribute to the whole composite.

SUMMARY

The dielectric constant in CBPCs fabricated with Wollastonite powders and $BaTiO_3$ has been presented. Generally, it has been shown that as $BaTiO_3$ increased, the dielectric constant increased and the loss tangent decreased. Also, indicated was how mean compressive strength increased as the $BaTiO_3$ has increased. In addition, results showed that as $BaTiO_3$ content increases, the density increases.

X-ray revealed that after mixing the $BaTiO_3$ with the mixing of Wollastonite and the acidic formulation, no new crystalline phases appeared as was planned; however, some amorphous sub-products could appear decreasing the impact of the $BaTiO_3$ dielectric properties in the ceramic composite. More research is being conducted on this topic.

ACKNOWLEDGEMENTS
The authors wish to thank to Colciencias-LASPAU for a grant to Henry A. Colorado and Composites Support and Solutions for supplying raw materials.

REFERENCES
[1] Della M. Roy. New Strong Cement Materials: Chemically Bonded Ceramics. February 1987 Science, Vol. 235 651

[2] Wilson, A. D. and Nicholson, J. W. Acid based cements: their biomedical and industrial applications. Cambridge, England, Cambridge University Press, 1993.

[3] Arun S. Wagh and Seung Y. Jeong. Chemically bonded phosphate ceramics: I, a dissolution model of formation. J. Am. Ceram. Soc., 86 [11] 1838-44 (2003).

[4] L. C. Chow and E. D. Eanes. Octacalcium phosphate. Monographs in oral science, vol 18. Karger, Switzerland, 2001.

[5] D. Singh, S. Y. Jeong, K. Dwyer and T. Abesadze. Ceramicrete: a novel ceramic packaging system for spent-fuel transport and storage. Argonne National Laboratory.Proceedings of Waste Management 2K Conference, Tucson, AZ, 2000.

[6] H. A. Colorado, C. Hiel and H. T. Hahn, J. M. Yang, J. Pleitt, C. Castano. Wollastonite-based Chemical Bonded Phosphate Ceramic with lead oxide contents under gamma irradiation. Journal of Nuclear Materials, 2011, DOI: 10.1016/j.jnucmat.2011.08.043

[7] T. L. Laufenberg, M. Aro, A. Wagh, J. E. Winandy, P. Donahue, S. Weitner and J. Aue. Phosphate-bonded ceramic-wood composites. Ninth International Conference on Inorganic bonded composite materials (2004).

[8] H. A. Colorado, C. Hiel and H. T. Hahn. Chemically Bonded Phosphate Ceramics composites reinforced with graphite nanoplatelets. Composites Part A. Composites: Part A 42 (2011) 376–384.

[9] Environmental Protection Agency AP 42 - Compilation of Air Pollutant Emission Factors, 2005 Volume I Stationary Point and Area Sources, Arunington, DC.

[10] J. F. Young and S. Dimitry. Electrical properties of chemical bonded ceramic insulators. J. Am. Ceram. Soc., 73, 9, 2775-78 (1990).

[11] X. Zhang, X. Z. Ding, C. K. Ong, B. T. G. Tan. Dielectric and electrical properties of ordinary Portland cement and slang cement in the early hydration period. Journal of Materials Science 31 (1996) 1345-1352.

[12] H. A. Colorado, C. Hiel and H. T. Hahn. Processing-Structure-Property Relations of Chemically Bonded Phosphate Ceramic Composites. Bulletin of Materials Science (Springer) Vol. 34, Suppl. No. 1. March 2011, pp 1-8.

[13] H. A. Colorado, C. Hiel, H. T. Hahn, Effects of particle size distribution of Wollastonite on curing and mechanical properties of the chemically bonded phosphate ceramics, in: N. Bansal, J. Singh, J. Lamon, S. Choi. (Eds.), Advances in Ceramic Matrix Composites, Wiley: Ceramic Transactions, New Jersey, 2011, pp. 85-98. Hard cover. ISSN: 10421122.

[15] Pragya Pandit, S. Satapathy, poorva Sharma, P. K. Gupta, S. M. Yusuf and V. G. Sathe. Structural, dielectric and multiferroic properties of Er and La substituted BiFeO3 ceramics. Bull. Mater. Sci. Vol. 34, No. 4, July 2011, pp 899-905.

# EQUIVALENT CIRCUIT MODELING OF CORE-SHELL STRUCTURED CERAMIC MATERIALS

Andreja Eršte,* Barbara Malič, Brigita Kužnik, Marija Kosec, and Vid Bobnar
Jožef Stefan Institute, Jamova cesta 39
Ljubljana, Slovenia

## ABSTRACT

The dielectric response of core-shell structured ceramic material has been modelled in terms of an equivalent electric circuit with elements that describe distinctive contributions of grains and grain boundaries. By taking into account a proper temperature dependence of individual elements of the circuit, the temperature and frequency dependent dispersive dielectric behavior, typically observed in these materials, has been obtained. The equivalent circuit modelling has been applied to the experimentally detected dielectric response of $CaCu_3Ti_4O_{12}$ thin films. In addition, the influence of parasitic inductance and resistance of the measuring setup on measured dielectric response will be presented.

## INTRODUCTION

Over the last few decades, heterogeneous ceramic materials, such as $CaCu_3Ti_4O_{12}$ (CCTO) bulk, thick or thin film,[1-3] $Gd_{0.6}Y_{0.4}BaCo_2O_{5.5}$,[4] and $(Sr,La)NbO_{3.5-x}$[5] have been intensively studied due to very high values of the dielectric constant that reveal immense potential for the use of these materials in various modern electronic and electromechanical applications. Dielectric spectroscopy turned out to be a powerful tool for revealing the electrical heterogeneities in the microstructure, i.e., core–shell structure, as the origin of such high values, which are almost constant in a broad frequency and temperature range.[4,6,7] A straightforward approach to understanding the dielectric behavior is analysis in terms of the equivalent circuit with elements that describe distinctive contributions to the dielectric response, i.e., analysis in terms of an equivalent circuit enables separation and characterization of distinctive contributions to the dielectric response.[2,8]

Figure 1a shows a simplified equivalent circuit, composed of two parallel RC elements connected in series. Each RC element (leaky capacitor) describes one contribution to the dielectric response of the material, e.g., in the case of CCTO single crystal one leaky capacitor describes the contributions of the bulk and the other one the contributions of surface layers.[7] In order to account for semiconduction of bulk material or influence of surface layers, equivalent circuit can be modified by introduction of new elements: Figure 1b shows a model for the case in which three distinctive contributions govern the dielectric response: grains, grain boundaries, and surface layers[1] and Figure 1c depicts the case of core–shell structured material, which is composed of semiconducting grains and insulating grain boundaries. As usual in semiconductors,[4] intrinsic effect of bulk sample is given as a sum of (i) frequency dependent ac conductivity, for which the universal dielectric response (UDR) is used with the ansatz $\sigma_{AC} = \sigma_0\omega^s$, $s < 1$, (ii) dc conductivity, and (iii) high-frequency dielectric constant. The intrinsic complex conductivity ($\sigma^* = \sigma' + i\sigma''$) is thus given as

$$\sigma' = \sigma_{DC} + \sigma_0\omega^s$$
$$\sigma'' = \omega\varepsilon_0\varepsilon' = \tan\frac{s\pi}{2}\sigma_0\omega^s + \omega\varepsilon_0\varepsilon_\infty \ . \tag{1}$$

First, the focus of this work will be modelling the temperature and frequency dependent dielectric response of core–shell structured materials. Then, experimentally detected response of CCTO thin films will be presented. In the end, the use of equivalent circuit for taking parasitic inductance and resistance of the measuring setup into account will be shown, i.e., how to rescue measurement data, compromised by the measuring setup.

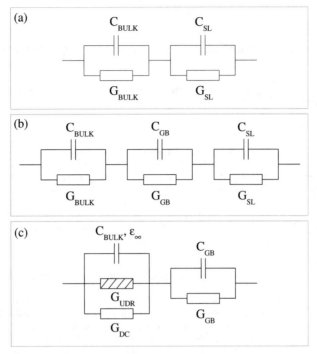

Figure 1. Equivalent circuits used for modelling and analysis of dielectric response of heterogeneous materials. Elements describe distinctive contributions of bulk material ($G_{BULK}C_{BULK}$), surface layers ($G_{SL}C_{SL}$), insulating grain boundaries ($G_{GB}C_{GB}$), or semiconducting grains ($G_{DC}G_{UDR}\varepsilon_\infty$) to the dielectric response.

## MODELLING THE DIELECTRIC RESPONSE OF CORE–SHELL STRUCTURED MATERIAL

In the simplest case of core–shell structured ceramic materials two constituents govern the dielectric response – conducting grains and insulating grain boundaries. This is similar to the case of single crystals (e.g., CCTO), where the non-intrinsic origin of colossal values of dielectric permittivity is due to crystal structure that consists of bulk material and surface layers, i.e., Schottky diode that was formed at the interface semiconductor-metal. Equivalent circuit used for analysis of such systems is composed of two leaky capacitors connected in series, as shown in Figure 1a: $G_{BULK}$ and $G_{SL}$ represent the conductance and $C_{BULK}$ and $C_{SL}$ the capacitance of bulk material and surface layers, respectively, and $C_{SL} \gg C_{BULK}$. The complex impedance of such equivalent circuit is

$$\frac{1}{Z} = \left[ \frac{1}{G_{BULK} + i\omega C_{BULK}} + \frac{1}{G_{SL} + i\omega C_{SL}} \right]^{-1}$$
$$= \frac{[(G_{BULK} + i\omega C_{BULK})(G_{SL} + i\omega C_{SL})][G_{BULK} + G_{SL} - i\omega(C_{BULK} + C_{SL})]}{(G_{BULK} + G_{SL})^2 + \omega^2(C_{BULK} + C_{SL})^2} \qquad (2)$$

and can be rewritten to form

$$\frac{1}{Z} = G_{MEAS} + i\omega C_{MEAS} \qquad (3)$$

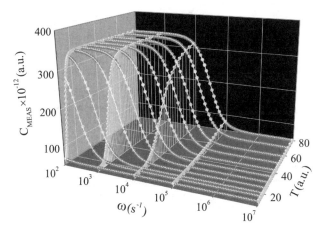

Figure 2. Modelled plot of $C_{MEAS}$ from Equation 3 vs frequency and temperature. Solid lines are a guide for the eye and, due to clarity, data is displayed in speed mode.

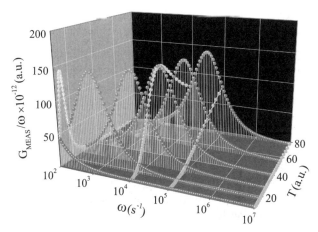

Figure 3. Modelled plot of $G_{MEAS}/\omega$ from Equation 3 vs frequency and temperature. Solid lines are a guide for the eye and, due to clarity, data is displayed in speed mode.

as these two quantities are usually measured by impedance analyzers in Cp-Gp mode.

Modelled plots of $C_{MEAS}$ and $G_{MEAS}/\omega$ from Equation 3 vs frequency and temperature are depicted in Figures 2 and 3, respectively, revealing an artificial Debye relaxation. The choice of parameters $G_{BULK}$, $C_{BULK}$, $G_{SL}$, and $C_{SL}$ for modelling does not influence the obtained Debye-like behavior as long as $C_{BULK} \ll C_{SL}$ and $G_{BULK} \gg G_{SL}$. In our case, some ordinary values have been used: $C_{BULK} = 50$ pF, $G_{BULK} = 10$ $\mu$S, $C_{SL} = 5$ nF, $G_{SL} = 0.001$ nS.

Characteristic frequency of the artificial Debye relaxation, derived from Equation 2 at $G_{BULK} \gg G_{SL}$ and $C_{BULK} \ll C_{SL}$, is

$$\omega \simeq 1/(C_{SL}R_{BULK}) , \tag{4}$$

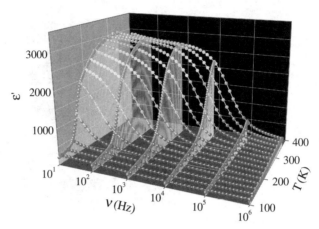

Figure 4. Temperature and frequency dependent $\varepsilon'$ of CCTO thin film. Solid lines are a guide for the eye and, due to clarity, data is displayed in speed mode.

where $R_{BULK} = 1/G_{BULK}$ represents the resistance of the bulk and $C_{SL}$ the capacitance of the electrode-bulk interface or grain boundaries. If $R_{BULK}$ and, concomitantly, the characteristic frequency of this artificial Debye's response follow the Arrhenius law, then, with increasing temperature giant surface layer/grain boundary contributions are detected up to higher frequencies. Thus, the temperature dependence of the conductivity of the bulk in Equation 2 is assumed to be the usual Arrhenius expression

$$1/G_{BULK} = R_{BULK} = R_0 \exp\left(-T/T_0\right) . \tag{5}$$

As the result, temperature dependencies of $C_{MEAS}$ and $G_{MEAS}/\omega$ in Figures 2 and 3 reveal an artificial Debye relaxation: At a fixed value of frequency the measured capacitance drops from $C_{SL}$ at higher temperatures to $C_{BULK}$ at lower temperatures, and, concomitantly, the characteristic drop temperature increases with frequency, which is indeed the experimental case reported in core–shell structured ceramic materials.[2,4,5]

EXPERIMENTALLY DETECTED DATA: CCTO THIN FILMS

CCTO thin film, prepared by chemical solution deposition,[3] with sample thickness of 540 nm and sputtered upper Cr/Au electrodes with 0.4 mm diameter was used for dielectric measurements. The complex dielectric constant $\varepsilon^*(\nu, T) = \varepsilon' - i\varepsilon''$ was measured by Novocontrol Alpha High Resolution Dielectric Analyzer in a frequency range of 1 Hz – 3 MHz. The amplitude of the probing ac electric signal was 50 mV. The data were obtained in the temperature range of 90 – 410 K during cooling with temperature change rate of $\Delta T/\Delta t = 0.5$ K/min. Temperature of samples was stabilized within 10 mK using a lock-in bridge technique with a Pt100 resistor as a thermometer.

Figures 4 and 5 show temperature and frequency dependence of $\varepsilon'$ and $\varepsilon''$, respectively. Comparison of measured dielectric data in Figures 4 and 5 to modelled dielectric behavior in Figures 1 and 2 in both cases reveals artificial Debye relaxation with characteristic relaxation frequency, which increases on increasing temperature.

For analysis of the detected dielectric response, equivalent circuit from Figure 1c has been applied as the material is composed of semiconducting grains and insulating grain boundaries. Although surface layers form at the electrode-sample interface in CCTO ceramic sample,[2] their contributions are not taken into account in this case. The analysis thus revealed values $C_{GB}$

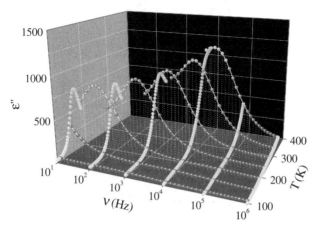

Figure 5. Temperature and frequency dependent $\varepsilon''$ of CCTO thin film. Dashed lines are a guide for the eye and, due to clarity, data is displayed in speed mode.

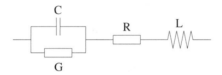

Figure 6. Equivalent circuit composed of leaky capacitor (sample), resistor (resistance of measuring setup), and inductor (inductance of measuring setup).

and $G_{GB}$, as well as of $C_{BULK}$, $G_{DC}$, and UDR parameter $s$,[3] and shows that high value of the dielectric constant arises due to the electrical conductivity of both grains and grain boundaries and, concomitantly, their influences on dielectric constant via Kramers–Kronig relations.

## EQUIVALENT CIRCUIT: RESONANCE AND RESISTANCE OF THE MEASURING SETUP

Experimentally detected dielectric data could be compromised due to the parasitic inductance or resistance of the measuring setup. Both, inductance and resistance of the measuring setup can be taken into account by preforming analysis in terms of the equivalent circuit[10] composed of a resistor and/or an inductor connected in series to a leaky capacitor (in impedance spectroscopy, intrinsic effect of bulk sample is usually presented by a leaky capacitor).

Figure 6 shows equivalent circuit for case, where both resistance and impedance of the measuring setup influence detected dielectric response. Complex impedance of such equivalent circuit is

$$\frac{1}{Z} = \left[\frac{1}{G + i\omega C} + R + i\omega L\right]^{-1} . \tag{6}$$

If we rewrite this term to form from Equation 3, we obtain

$$G_{MEAS} = \frac{G + R(G^2 + \omega^2 C^2)}{(1 - \omega^2 LC + GR)^2 + (\omega LG + \omega CR)^2}$$

$$C_{MEAS} = \frac{C - L(G^2 + \omega^2 C^2)}{(1 - \omega^2 LC + GR)^2 + (\omega LG + \omega CR)^2} . \tag{7}$$

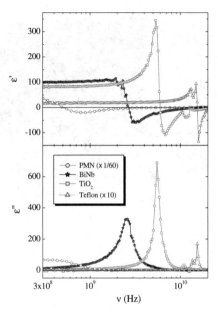

Figure 7. Resonance due to the measuring setup in various samples, measured with 8720C Network Analyzer.

At $\omega^2 LC = 1$ resonance occurs due to inductance of the measuring setup. This can clearly be seen in Figure 7, which depicts resonance due to the measuring setup for several samples at room temperature. Note, that the frequency of the resonance $\omega = 1/\sqrt{LC}$ depends on material used for measurements – higher is the dielectric constant, lower is the frequency.

If measurements are preformed in frequency range, which is low enough to avoid resonance due to the measuring setup, then $L = 0$ can be assumed. In this case, the result of Equation 6 reads as

$$G_{MEAS} = \frac{G + R(G^2 + \omega^2 C^2)}{(1 + GR)^2 + (\omega CR)^2}$$
$$C_{MEAS} = \frac{C}{(1 + GR)^2 + (\omega CR)^2} \cdot$$

(8)

Even a small resistance of the measuring setup (a few Ohm – in this case the term $GR$ in the denominator of Equation 8 can be neglected) results in an artificial Debye-like relaxation with a frequency of $\omega \simeq \sqrt{2}/CR$ in the detected dielectric response. Although this relaxation frequency, above which the response is incorrect, is usually very high ($C = 50$ pF and $R = 3\ \Omega$ give $\nu = \omega/2\pi \simeq 1.5$ GHz), it can decrease into the measuring frequency window when samples with high capacity are investigated.

CONCLUSION

Temperature and frequency dependent dielectric response of core–shell structured ceramic material has been modelled in terms of the equivalent circuit by assuming the Arrhenius law for temperature dependance of bulk resistance. Modelling results reveal a dispersive dielectric behavior, typically observed in core–shell structured material, e.g., CCTO thin film.

The use of analysis in terms of the equivalent circuit as a data rescue tool has been presented for cases in which detected dielectric data are compromised by resistance or resonance of the measuring setup. In addition, it has been shown that even small values of resistance or inductance result in an artificial relaxation or resonance in the detected dielectric response.

FOOTNOTES

*Corresponding author, e–mail: andreja.erste@ijs.si

REFERENCES

[1] T. B. Adams, D. C. Sinclair, and A. R. West, Characterization of Grain Boundary Impedances in Fine- and Coarse-Grained $CaCu_3Ti_4O_{12}$ Ceramics, *Phys. Rev. B*, **73**, 094124 (2006).

[2] P. Lunkenheimer, S. Khrons, S. Riegg, S. G. Ebbighaus, A. Reller, and A. Loidl, Colossal Dielectric Constants in Transition-Metal Oxides, *Eur. Phys. J. Special Topics*, **180**, 61–89 (2010).

[3] A. Eršte, B. Malič, B. Kužnik, M. Kosec, and V. Bobnar, Influence of Preparation Conditions on Distinctive Contributions to Dielectric Behavior of $CaCu_3Ti_4O_{12}$ Thin Films, *J. Am. Ceram. Soc.*, doi: 10.1111/j.1551-2916.2011.04581.x (in press)

[4] V. Bobnar, P. Lunkenheimer, M. Paraskevopoulos, and A. Loidl, Separation of Grain Boundary Effects and Intrinsic Properties in Perovskite-Like $Gd_{0.6}Y_{0.4}BaCo_2O_{5.5}$ Using High-Frequency Dielectric Spectroscopy, *Phys. Rev. B*, **65**, 184403 (2002).

[5] V. Bobnar, P. Lunkenheimer, J. Hemberger, A. Loidl, F. Lichtenberg, J. Mannhart, Dielectric Properties and Charge Transport in the $(Sr,La)NbO_{3.5-x}$ System, *Phys. Rev. B*, **65**, 155115 (2002).

[6] M. A. Subramanian, D. Li, N. Duan, B. A. Reisner, and A. W. Sleight, High Cielectric Constant in $ACu_3Ti_4O_{12}$ and $ACu_3Ti_3FeO_{12}$ Phases, *J. Solid State Chem.*, **151**, 323–325 (2000).

[7] P. Lunkenheimer, R. Fichtl, S. G. Ebbinghaus, and A. Loidl, Nonintrinsic Origin of the Colossal Dielectric Constants in $CaCu_3Ti_4O_{12}$, *Phys. Rev. B*, **70**, 172102 (2004).

[8] D. C. Sinclair, T. B. Adams, F. D. Morrison, and A. R. West, $CaCu_3Ti_4O_{12}$: One-Step Internal Barrier Layer Capacitor, *Appl. Phys. Lett.*, **80**, 2153-55 (2002).

[9] P. Lunkenheimer, V. Bobnar, A. V. Pronin, A. I. Ritus, A. A. Volkov, and A. Loidl, Origin of Apparent Colossal Dielectric Constants, *Phys. Rev. B*, **66**, 052105 (2002).

[10] F. Kremer and A. Schönhals, Broadband Dielectric Spectroscopy, Berlin: Springer, 2003.

# Bi$_2$Te$_3$ AND Bi$_2$Te$_{3-x}$S$_x$ FOR THERMOELECTRIC APPLICATIONS

[1]W. Wong-Ng, [2]N. Lowhorn, [1]J. Martin, [3]P. Zavalij, [4]H. Joress, [1]Q. Huang, [1]Y. Yan, [5]A. N. Mansour, [6]E. L. Thomas, [7]J. Yang, and [1]M. L. Green

[1] NIST, Gaithersburg, MD 20899
[2] Physics Department, Eastern State University of Tennessee, TN 37128
[3] Chemistry Department, University of Maryland, College Park, MD 20742
[4] Materials Science and Engineering Department, Johns Hopkins University, Baltimore, MD 21218
[5] NSWC, Carderock Division, West Bethesda, MD 20817
[6] University of Dayton Research Institute Air Force Research Laboratory, WPAFB, OH 45433
[7] Materials Science and Engineering Department, University of Washington, Seattle, WA 98105

ABSTRACT
    The increased interest in research and development of thermoelectric materials is partly due to the need for improved efficiency in the global utilization of energy resources. Historically, Bi$_2$Te$_3$ occupies an important position in the thermoelectric community. For commercial cooling applications, Bi$_2$Te$_3$ is currently the most practical and efficient material. Compounds in the Bi$_2$Te$_{3-x}$S$_x$ system have also been studied in effort to further improve thermoelectric performance. This paper summarizes our two recent research efforts on the bismuth telluride systems, including the certification of a standard reference material (SRM$^{\circledR}$) for low temperature Seebeck coefficient using Te-doped Bi$_2$Te$_3$, and the composition/structure/property relationship studies of the effect of S alloying with Te in a Bi$_2$Te$_{3-x}$S$_x$ single crystal.

## I. INTRODUCTION

    Thermoelectric materials have demonstrated potential for both waste heat recovery and solid-state refrigeration applications. The efficiency and performance of thermoelectric materials for power generation and cooling is related to the dimensionless figure of merit (ZT) of the thermoelectric materials, given by $ZT = S^2\sigma T/\kappa$, where $T$ is the absolute temperature, $S$ is the Seebeck coefficient, $\sigma$ is the electrical conductivity ($\sigma = 1/\rho$, $\rho$ is electrical resistivity), and $\kappa$ is the thermal conductivity [1, 2]. $ZT$ is directly related to the performance of a thermoelectric material and is the reference by which these materials are judged. The Seebeck coefficient is an important indicator for power conversion efficiency. Thermoelectric materials with desirable properties (i.e., high $ZT \gg 1$) are characterized by high electrical conductivity, high Seebeck coefficient, and low thermal conductivity and will have widespread applications. Until recently, only a small number of materials have been found to have practical industrial applications because of their generally low thermoelectric efficiencies.

    Bismuth telluride, Bi$_2$Te$_3$, is a classic narrow-band-gap semiconductor that has been used extensively in cooling applications because of its relatively large Seebeck coefficient [3]. The $ZT$ value for Bi$_2$Te$_3$ is about 1 at room temperature. In order to improve the performance of this material, much work has been carried out using dopants either on the Bi-site or on the Te-site.

    The goal of this paper is to summarize our recent efforts on the study of two systems based on Bi$_2$Te$_3$. The first effort pertains to a study of the structure/property relation in a bismuth telluride sulfide single crystal, Bi$_2$Te$_{3-x}$S$_x$. Techniques of characterization included non-ambient neutron and X-ray diffraction and EXAFS, and a set of thermoelectric property measurement tools for Seebeck coefficient and carrier concentration. The second effort is the development of a low temperature Seebeck coefficient standard reference material (SRM$^{®}$), using a Te-doped Bi$_2$Te$_3$ material. The availability of standard reference materials will validate measurement accuracy, leading to a better understanding of the structure/property relationships and the underlying physics of new and improved thermoelectric materials. Therefore, the development of an SRM$^{®}$ for thermoelectric research is critical for the advancement of U.S. industry.

## II. STRUCTURE/PROPERTY STUDIES OF Bi$_2$Te$_{3-x}$S$_x$ CRYSTALS

Single crystal specimens are useful for measurement of anisotropic thermoelectric properties. It is also desirable to have information on the homogeneity of the crystals, as this affects the properties in different regions of the crystal. The bulk Bi$_2$Te$_{3-x}$S$_x$ crystals were prepared by induction melting technique [4]. Large crystals with dimensions in the centimeters range were obtained. One relatively large crystal with slightly irregular shape ($\sim$ 0.4 cm x 0.1 cm at the wide end, and 0.2 cm x 0.1 cm at the narrow end and 4.5 cm in length (Fig. 1)), was selected for this investigation.

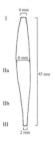

Fig. 1. Bi$_2$Te$_{3-x}$S$_x$ crystal with labels of the locations where the three small single crystals (locations I, IIa, and III) were cut for the structural studies, and the three slices were cut (I, IIb, and III) for Hall measurements after the screening experiments were completed.

A high-throughput Seebeck coefficient screening tool, developed at National Institute of Standards and Technology (NIST), was used for measurement along the long axis of the crystal [5, 6, 7]. The salient features of the screening tool consist of a probe to measure Seebeck coefficient, an automated translation stage to scan the sample in the $x$-$y$-$z$ directions, and various voltage measuring instruments. The details of the tool have been described elsewhere [5, 6, 7]. All Seebeck coefficient measurements were conducted at room temperature with $\Delta T = 4.4$ K. We also determined the carrier concentration on corresponding areas of the crystal (locations I, IIb, III) using a Hall measurements. The Hall measurements were conducted at 300 K on representative bar-shaped samples (with approximate dimensions 0.5 mm x 1 mm x 4 mm, cut from the crystal) using the Physical Property Measurement System (PPMS, Quantum Design)[1] infrastructure and a custom measurement sequence.

To obtain the structure of Bi$_2$Te$_{3-x}$S$_x$ as a function of distance along the long-direction of the crystal, three small pieces of crystals ranging from 0.01 mm to 0.3 mm were cut from the two ends and the middle part of the crystal along the 4.5 cm axis for single crystal x-ray crystallographic analysis. Figure 1 also gives the schematic of the Bi$_2$Te$_{3-x}$S$_x$ crystal illustrating the locations (I, IIa, and III) where the small crystals were cut for structure characterization. X-ray intensity data were measured at 200K on a three-circle diffractometer system equipped with Bruker Smart Apex II CCD area detector using a graphite monochromator and a MoKα fine-focus sealed tube ($\lambda= 0.71073$ Å)[1].

The Bi$_2$Te$_{3-x}$S$_x$ crystal was found to have a concentration gradient along the long axis, as deduced from the crystallographic data (Table 1). Te has the highest concentration in location I (Bi$_2$Te$_{1.58(1)}$S$_{1.42(1)}$), followed by location IIa (Bi$_2$Te$_{1.53(1)}$S$_{1.47(1)}$), and the least in location III (Bi$_2$Te$_{1.52(2)}$S$_{1.48(2)}$). As the covalent radius of Te is larger than that of S [8], the greater content of Te in location I yields the largest cell volume among the three sites, as expected (Fig 2). Bi$_2$Te$_3$ can be visualized in terms of a 2-dimensional layered structure (Fig 3) [9]. The structure consists of repeated quintuple layers of atoms (Te1-Bi-Te2-Bi-Te1) stacking along the $c$-axis of the unit cell. The Te2 site in the inner quintuple layers of Bi$_2$Te$_3$ was fully occupied by S, whereas in the outer van der Waals gap Te1 layer, an approximately 21 % (location I) to 24 % (location III) of the Te1 site is substituted by S.

Table 1. Crystallographic Data for Bi$_2$S$_{1.42(1)}$Te$_{1.58(1)}$, Bi$_2$S$_{1.47(1)}$Te$_{1.53(1)}$, and Bi$_2$S$_{1.48(2)}$Te$_{1.52(2)}$ (space group $R$-$3m$, and $Z$=3).

| Chemical formula | Bi$_2$S$_{1.42(1)}$Te$_{1.58(1)}$ (I) | Bi$_2$S$_{1.47(1)}$Te$_{1.53(1)}$ (IIa) | Bi$_2$S$_{1.48(2)}$Te$_{1.52(2)}$ (III) |
|---|---|---|---|
| Crystal size | 0.01 × 0.14 × 0.34 mm | 0.03 × 0.09 × 0.27 mm | 0.04 × 0.08 × 0.29 mm |
| Unit cell dimensions | $a$ = 4.1968(4) Å | $a$ = 4.1900(5) Å | $a$ = 4.1879(7) Å |
| | $c$ = 29.467(6) Å | $c$ = 29.445(7) Å | $c$ = 29.405(10) Å |
| | $V$=449.47(11) Å$^3$ | $V$=447.68(13) Å$^3$ | $V$=446.6 (2) Å$^3$ |

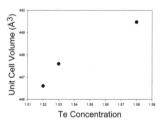

Fig. 2. Plot of unit cell volume ($V$) vs concentration of 'Te'

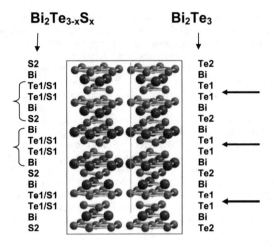

Fig. 3. Crystal Structure of Bi$_2$Te$_{3-x}$S$_x$ showing the groups of quintuple-layer of Te1/S1-Bi-S-Bi-Te1/S1. The position of the van der Waals planes (arrows) are illustrated.

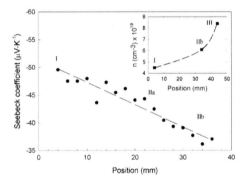

Fig. 4. Seebeck coefficient and carrier concentration of the Bi$_2$Te$_{3-x}$S$_x$ crystal as a function of the distance along the 4.5 cm axis.

Figure 4 plots the Seebeck coefficients of the 16 data points and the carrier concentration of the 3 samples in locations I, IIb, and III, as a function of the distance from the wide end of the crystal. The composition gradient along the long-axis of the crystal correlates well with the Seebeck coefficient values and the carrier concentration. The greater the S concentration (resulting in a decrease in the hole concentration), the lower the Seebeck coefficient and the higher the carrier density of Bi$_2$Te$_{3-x}$S$_x$, as expected from a rigid band semiconductor model. We have extended the versatility of the screening tool from application for thin film to single crystals.

## III. LOW TEMPERATURE SEEBECK COEFFICIENT SRM® (10K to 390K)

SRMs® are available for thermal conductivity and electrical conductivity (NIST SRM® 8420/8421-electrolytic iron and 8424/8426-graphite), however, a moderate to high Seebeck coefficient standard reference material does not exist. To enable inter-laboratory data comparison and validation of measurements, we have initiated a project to develop Seebeck coefficient SRMs® for both low temperature and high temperature applications. Prior to this development, we have completed the certification of a low temperature Seebeck coefficient standard reference material (SRM 3451). A round-robin measurement survey was first conducted to determine the appropriate material and to examine different measurement techniques. Two candidate materials, constantan and undoped-Bi$_2$Te$_3$ were circulated between 12 laboratories actively involved in thermoelectric research. The details and results from this survey are presented elsewhere [10, 11]. As a result of the round robin study, Bi$_2$Te$_3$ was chosen as the prototype SRM.

The Low Temperature Seebeck Coefficient Standard Reference Material, SRM 3451, is a bar-shaped artifact of Te-doped Bi$_2$Te$_3$ [(Bi$_{1.998}$Te$_{0.002}$)Te$_3$] (approximately 3.5 mm x 2.5 mm x 8.0 mm) of non-stoichiometric bismuth telluride (n-type, Te rich, Bi/Te ratio approximately 2/3, formula Bi$_2$Te$_{3+x}$) (Fig. 5). The SRM® artifacts were custom prepared (involving ingot solidification, cutting and polishing of the bar-shaped pieces, and metal coating of their ends) for NIST by Marlow Industries, Inc.

Fig. 5. Photograph of the SRM 3451 artifact.

The Seebeck coefficient is defined as the ratio of the Seebeck voltage ($\Delta V$) to the applied temperature gradient ($\Delta T$) [12]. The basic concept of a typical Seebeck measurement requires the creation of a $\Delta T$, then measurement of the $\Delta T$ and resulting $\Delta V$. For our certification measurements, we used a custom (with third party instruments), steady-state, multiple gradient technique using a modified Quantum Design PPMS (Model PPMS-9). In this technique, the sample was held at a constant temperature while a range of $\Delta T$ values were created and the corresponding $\Delta V$ values measured. Graphing this data produced a line from which the slope yielded the Seebeck coefficient. Custom software was written using LabVIEW [13] to automate the measurements and data collection.

Ten samples were selected randomly from the SRM batch of 390 received from Marlow Industries. The details of the sample mounting and calibration of thermometers are discussed elsewhere [14]. Five of the samples were measured twice using the primary technique. The other five samples were measured once with this same technique. The Seebeck coefficient values (in units of $\mu V/K$) were measured at 32 temperature values for each of the 10 samples. Plot of NIST certified Seebeck coefficient data with 95 % confidence band is shown on Fig. 6. The certified Seebeck coefficient values ($S_m$), type A (random error, $\sigma_A$) and type B (systematic error, $\sigma_B$) uncertainty components, the combined standard uncertainty ($\sigma$), the expanded uncertainty, and coefficient of variation ($\sigma$/mean) for SRM 3451 are listed in Table 2.

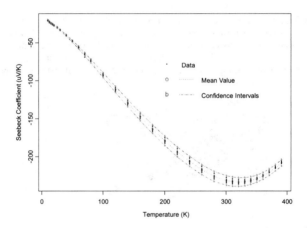

Fig. 6. Plot of certified Seebeck coefficient data with 95 % confidence band. Dotted lines are guides for the eye.

Table 2: Certified Seebeck coefficient values for SRM 3451.

| Temp (K) | Seebeck Coefficient, $S_m$ ($\mu$V/K) | $\sigma_A$ | $\sigma_B$ | $\sigma$ | Expanded Uncertainty (Coverage factor = 2) | Coefficient of variation |
|---|---|---|---|---|---|---|
| 10.09 | -20.83 | 0.39 | 0.18 | 0.43 | 0.85 | 0.021 |
| 12.58 | -22.98 | 0.47 | 0.2 | 0.52 | 1.03 | 0.022 |
| 15.1 | -24.56 | 0.41 | 0.22 | 0.47 | 0.93 | 0.019 |
| 17.6 | -25.95 | 0.42 | 0.23 | 0.48 | 0.96 | 0.018 |
| 20.09 | -27.3 | 0.25 | 0.24 | 0.35 | 0.71 | 0.013 |
| 25.1 | -29.79 | 0.35 | 0.27 | 0.44 | 0.88 | 0.015 |
| 30.11 | -33.36 | 0.17 | 0.30 | 0.35 | 0.7 | 0.010 |
| 40.14 | -40.42 | 0.27 | 0.37 | 0.46 | 0.92 | 0.011 |
| 50.16 | -48.01 | 0.39 | 0.44 | 0.59 | 1.18 | 0.012 |
| 60.25 | -56.26 | 0.60 | 0.52 | 0.80 | 1.60 | 0.014 |
| 70.29 | -65.28 | 0.93 | 0.61 | 1.11 | 2.22 | 0.017 |
| 80.33 | -74.1 | 0.94 | 0.69 | 1.17 | 2.34 | 0.016 |
| 100.34 | -92.79 | 1.30 | 0.88 | 1.57 | 3.13 | 0.017 |
| 120.37 | -111.6 | 1.69 | 1.06 | 1.99 | 3.99 | 0.018 |
| 140.38 | -129.6 | 1.95 | 1.23 | 2.31 | 4.61 | 0.018 |
| 160.4 | -147.22 | 2.21 | 1.39 | 2.61 | 5.22 | 0.018 |
| 180.41 | -163.81 | 2.41 | 1.53 | 2.85 | 5.70 | 0.017 |
| 200.43 | -179.4 | 2.61 | 1.65 | 3.09 | 6.19 | 0.017 |
| 220.49 | -193.96 | 2.71 | 1.76 | 3.23 | 6.46 | 0.017 |
| 240.52 | -206.22 | 2.73 | 1.84 | 3.03 | 6.59 | 0.016 |
| 260.52 | -217.26 | 2.73 | 1.91 | 3.33 | 6.66 | 0.015 |
| 280.71 | -226.11 | 2.64 | 1.95 | 3.28 | 6.56 | 0.015 |
| 300.73 | -231.36 | 2.43 | 1.98 | 3.13 | 6.26 | 0.014 |
| 310.74 | -232.82 | 2.27 | 1.98 | 3.01 | 6.03 | 0.013 |
| 320.74 | -233.48 | 2.08 | 1.98 | 2.87 | 5.75 | 0.012 |
| 330.87 | -232.99 | 1.81 | 1.98 | 2.68 | 5.37 | 0.012 |
| 340.84 | -231.45 | 1.55 | 1.98 | 2.51 | 5.02 | 0.011 |
| 350.81 | -228.93 | 1.26 | 1.97 | 2.34 | 4.67 | 0.010 |
| 360.76 | -225.28 | 1.03 | 1.96 | 2.21 | 4.42 | 0.010 |
| 370.9 | -220.43 | 0.80 | 1.94 | 2.10 | 4.21 | 0.010 |
| 381.04 | -214.58 | 0.79 | 1.93 | 2.08 | 4.16 | 0.010 |
| 391 | -208.17 | 0.90 | 1.92 | 2.12 | 4.23 | 0.010 |

In order for the user to obtain the Seebeck coefficient of the artifact at temperatures in between the thirty-two base temperature values at which it was measured, the thirty-two ($S_m$, $T$) data points (Table 2) can be fit to a fourth order polynomial function, expanded around a temperature A:

$$S_m(T) = S_A + aT\left(1 - \frac{A}{T}\right) + bT^2\left(1 - \frac{A}{T}\right)^2 + cT^3\left(1 - \frac{A}{T}\right)^3 + dT^4\left(1 - \frac{A}{T}\right)^4 \quad (1)$$

where $S_m(T)$ is the interpolated value of the Seebeck coefficient, and 10 K ≤ A ≤ 391 K. As an example, if A is chosen for utility to be room temperature (295 K), the following values are found for the fitting coefficients:

$a$ = -2.2040 x 10$^{-1}$ µV/K$^2$
$b$ = 3.9706 x 10$^{-3}$ µV/K$^3$
$c$ = 7.2922 x 10$^{-6}$ µV/K$^4$
$d$ = -1.0864 x 10$^{-9}$ µV/K$^5$

Further, $S_A$ = -230.03 µV/K, and $R^2$ = 0.99994649. Expansion of eq. (1) around other values of A led to insignificant differences in interpolated values over the entire temperature range. Fig. 7 is a plot of eq. 1 for A = 295 K. This SRM will be available in the Fall of 2011.

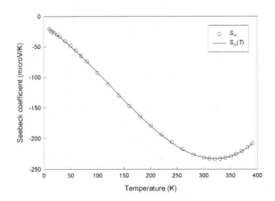

Fig. 7. Fourth order polynomial fit to the measured data for SRM 3451

We also report the electrical resistivity data as the supplementary reference (informational) data. The electrical resistivity was measured on six of the ten SRM artifacts using the AC transport option of the PPMS. A standard four point probe technique was used. Figure 8 gives the resistivity data as a function of temperature.

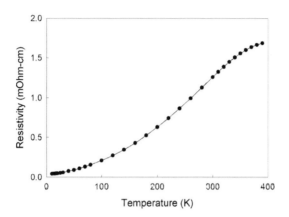

Fig. 8. SRM artifact resistivity as a function of temperature showing mean values and 95 % confidence intervals.

Additional characterization data of SRM3451 includes thermal expansion coefficient and the analysis of the (thermal) displacement parameters using neutron diffraction and EXAFS, respectively, as a function of temperature. From 20 K to 300 K, the unit cell volume increases by 1.2 %. The equation of volume expansion was obtained as $V = M_0 + M_1T + M_2T^2$, where $M_0$=501.12, $M_1$=0.017396, $M_2$=1.1072 x10$^{-5}$. Fourier transforms of temperature dependent Bi L$_3$-edge EXAFS spectra displaying contributions from the first few coordination spheres for Bi$_2$Te$_3$ were obtained at temperatures of (19, 50, 75, 100, 125, 150, 175, 200, 225, 250, 275, and 298) K, but only those taken at 19 K, 100 K, 200 K, and 298K are shown in Fig. 9 as examples. The amplitude of the Fourier transforms increased with decreasing temperatures due to quenching of thermal motion of the atoms. The Einstein temperatures $\theta_E$ (148(3), 121(3), and 79(2) [15] for the 1$^{st}$ (Bi-Te), 2$^{nd}$ (Bi-Te), and 3$^{rd}$ (Bi-Bi) shell, respectively) are consistent with soft phonon modes. The Einstein temperature for the Bi-Bi pair appears to be consistent with those of rattling atoms in cage-like structures [16] and is expected to enhance phonon-phonon scattering. Hence, the soft localized phonon modes are not restricted to cage-like structures as reported earlier in the case of zinc antimony [17].

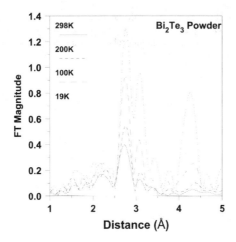

Fig. 9. Fourier transforms displaying contributions from the first few coordination spheres for Bi$_2$Te$_3$ are shown for T=19 K, 50 K, 75 K and 100 K.

IV. FUTURE PLAN

For applications in energy conversion industries including the automotive industry, it is important to have a high temperature Seebeck coefficient SRM. Currently we are in the process of completing a high temperature Seebeck coefficient measuring tool. After that, we plan to collaborate with the international thermoelectric research community to develop a high temperature Seebeck coefficient SRM.

We continue our effort on the improvement of our power factor ($S^2\sigma$, where $\sigma$ is electrical conductivity) screening tool for screening the thermoelectric combinatorial films. There are three limitations to the current screening tool configuration. First, it is difficult to measure materials with low (i.e., $<10^{-1}$ ($\Omega$cm)$^{-1}$ for a 200 nm thin film) electrical conductivity and low ($<5\mu$V/K) absolute value of Seebeck coefficient due to limitations of the current source and voltage meter used. Although these limitations do not hamper our ability to screen thermoelectric materials, we have extended our measurement capability by upgrading these two components. Second, the spatial resolution of our screening tool is about 2 mm, corresponding to the distance between two spring probes. If the electrical conductivity of the sample drastically changed over a very short length scale, measurements at a sample point might be affected by its neighbors. We are in the process of improving this resolution to about 1 mm. Third, measurements can be carried out only at room temperature at present. We are actively developing an improved instrument that is capable of high temperature (up to 800 °C) measurement.

ACKNOWLEDGMENT

ANM acknowledges financial support by the Carderock Division of the Naval Surface Warfare Center's In-house Laboratory Independent Research Program administrated under ONR's Program Element 0601152N. The XAS experiments were conducted at the National Synchrotron Light Source of Brookhaven National Laboratory, which is supported by the U.S. Department of Energy, Office of Basic Energy Sciences, under contract no. DE-AC02-98CH10886.

FOOTNOTE[1]

Certain commercial equipment, instruments, or materials are identified in this paper in order to specify the experimental procedure adequately. Such identification is not intended to imply recommendation or endorsement by the National Institute of Standards and Technology, nor is it intended to imply that the materials or equipment identified are necessarily the best available for the purpose.

REFERENCES

[1]H.J. Goldsmid, A.R. Sheard, D A. Wright, Brit. J. Appl. Phys. **9** 365 (1958).

[2]T.M. Tritt, and M.A. Subramanian, guest editors, *Harvesting Energy through Thermoelectrics: Power Generation and Cooling*, pp. 188-195, MRS Bulletin, Materials Research Soc. (2006).

[3]C. Drasar, P. Lostak, and C. Uher, J. Electronic Mater. **39**(9) 2162 (2010).

[4]H.H. Soonpaa and R.K. Mueller, General Mills, Inc., Mechanical Division Report No 2195 AD257897 (1961). [Bridgman]

[5]M. Otani, E. Thomas, W. Wong-Ng, P. K. Schenck, K.-S. Chang, N. D. Lowhorn, M. L. Green, and H. Ohguchi, A High-throughput Screening System for Thermoelectric Material Exploration Based on a Combinatorial Film Approach, Jap. J. Appl. Phys, **48**, 05EB02 (2009).

[6]M. Otani, N.D. Lowhorn, P.K. Schenck, W. Wong-Ng, and M. Green, A High-throughput thermoelectric Screening Tool for Rapid Construction of Thermoelectric Phase Diagrams, Appl. Phys. Lett., **91**, 132102-132104 (2007).

[7] M. Otani, K. Itaka, W. Wong-Ng, P.K. Schenck, and H. Koinuma, Development of high-throughput thermoelectric tool for combinatorial thin film libraries, Appl. Surface Science, **254**, 765-767 (2007).

[8]J. Horák, P. Lošták, L. Koudelka and R. Novotný, Solid State Communications, Vol. 55, No. 11, pp. 1031-1034 (1985).

[9]D. Harker, Z. Kristallogr. **89**, 175 (1934).

[10]N. D. Lowhorn, W. Wong-Ng, Z. Q. J. Lu, W. Zhang, E. Thomas, M. Otani, M. Green, T. N. Tran, C. Caylor, N. Dilley, A. Downey, B. Edwards, N. Elsner, S. Ghamaty, T. Hogan, Q. Jie, Q. Li, J. Martin, G. Nolas, H. Obara, J. Sharp, R. Venkatasubramanian, R. Willigan, J. Yang, and T. Tritt, Round-Robin Studies of Two Potential Seebeck Coefficient Standard Reference Materials, Appl. Phys. A, (Materials Science & processing), **94**, 231-234 (2009).

[11]Z. Q. J. Lu, N. D. Lowhorn, W. Wong-Ng, W. Zhang, E. Thomas, M. Otani, M. Green, T. N. Tran, C. Caylor, N. Dilley, A. Downey, B. Edwards, N. Elsner, S. Ghamaty, T. Hogan, Q. Jie, Q. Li, J. Martin, G. Nolas, H. Obara, J. Sharp, R. Venkatasubramanian, R. Willigan, J. Yang, and T. Tritt, Statistical Analysis of two Candidate Materials for a Seebeck Coefficient Standard Reference Material (SRM™), J. Res. Nat'l Inst. Stand & Tech., **114** (1), 37-55 (2009).

[12]J. Martin, T. Tritt, C. Uher, *High temperature Seebeck coefficient metrology*, J. Appl. Phys. **108**, 121101 (2010).

[13]LabVIEW is a programming language produced by National Instruments which is commonly used for experiment control.

[14]Lowhorn, N.D; Wong-Ng, W; Lu, Z.Q.; Martin, J.; Green; M.L.; Thomas, E.L.; Bonevich, J.E.; Dilley, N.R.; Sharp, J.; *Development of a Seebeck Coefficient Standard Reference Material (SRM)*; J. Mater. Res., **26** (15), pp.1983-1992 (2011).

[15]Rehr, J.J.; Albers, R.C.; *Theoretical Approaches to X-ray Absorption Fine Structure*; Rev. Mod. Phys. **72(3)**, 621 (2000)

[16]Baumbach, R.; Bridges, F.; Downward, L.; Cao, D.; Chesler, P.; Sales, B.; *Off-center phonon scattering sites in Eu$_8$Ga$_{16}$Ge$_{30}$ and Sr$_8$Ga$_{16}$Ge$_{30}$*, Phys. Rev. B, **71'** 024202 (2005)

[17]Schweika, W.; Hermann, R.P.; Prager, M.; Perβon, J., Keppens, V.; *Dumbbell Rattling in Thermoelectric Materials*; Phys. Rev. Lett., **99**, 125501-1 (2007)

# OPTIMIZED SPUTTERING PARAMETERS FOR ITO THIN FILMS OF HIGH CONDUCTIVITY AND TRANSPARENCY

Jihoon Jung and Ruyan Guo[*]
Department of Electrical and Computer Engineering
The University of Texas at San Antonio
San Antonio, TX 78249, USA

ABSTRACT
        Although many ways of deposition of ITO thin films are available, RF magnetron sputtering deposition is the most efficient way for high transparency, conductivity and reproducibility. In this work, ITO thin films were deposited on glass substrate by RF magnetron sputtering under a varying array of the sputtering parameters: RF power (30 to70W), working pressure (3 to 7 mTorr), oxygen partial pressure (0 to 2%), and substrate temperature (room temperature to 500 °C). Sputtering parameters such as the working pressure, RF power, oxygen partial pressure, and substrate temperature have been optimized for high conductivity and transparency. This paper reports correlations of the sputtering parameters with the properties of the deposited ITO films, using a wide range of characterization techniques including x-ray diffraction, atomic force microscope, four probe conductivity measurement and optical spectroscopy. The optimized conditions for high transmittance and low electrical resistivity using a low profile desktop sputtering tool are reported.

## INTRODUCTION

        Transparent conducting oxides (TCOs) exhibit both high electrical conduction and optical transparency. Transparent conducting oxides (TCOs) are vital components in many optoelectronic applications such as touch screens, flat panel displays and photovolatics [Thomas, 1997 and Wager, 2003]. Among many candidates for TCOs thin films, tin doped indium oxide (ITO) films have been extensively used for the applications of electronics and opto-electronics because of their high electrical conductivity, transparency and easiness of being deposited as a thin film[1]. It is known that ITO thin films are highly degenerate wide band gap (3.5-4.3 eV) n-type semiconductors resulted from oxygen vacancies and substitutional tin dopant[2].

        There are a variety of methods for a deposition of ITO films such as sputtering[3,4], thermal evaporation[5], and spray pyrolysis technique[6], but commercial ITO films are manufactured mostly by magnetron sputtering with an ITO ceramic target material[7] because of its large available deposition areas, high deposition rates, and less damaged areas[8] According to literature, the properties of ITO films deposited by RF magnetron sputtering is extensively dependant on the sputtering parameters such as RF power, working pressure, oxygen partial pressure, target to substrate distance, and substrate temperature[9].

        In this paper, we report on the trends of ITO films deposited on glass substrate by RF magnetron sputtering with different sputtering parameters. To optimize the best condition for the properties of the deposited ITO films, RF power, argon pressure, oxygen partial pressure, and substrate temperature were varied from 30W to 70W, from 3mTorr to 7mTorr, from 0% to 2%, and from room temperature to 500 C, respectively. The electrical, optical, and structural properties have been characterized as a function of sputtering parameters which are described

[*] corresponding email ruyan.guo@utsa.edu

above.

## EXPERIMENTAL DETAILS

The ITO films, having 50-200nm range, were deposited by commercial RF (13.56 MHz) magnetron sputtering system (Denton Vacuum, Inc., USA.). The target material employed was a ceramic target (90 wt % $In_2O_3$ and 10 wt% SnO, 99.99% purity) supplied by Angstrom Sciences, Inc., USA. The commercial soda lime glass (Thermo Scientific, Inc., USA) was used for the substrate and cleaned in acetone, ethanol, and de-ionized water in that order by an ultra sonic cleaner, then blown dry in nitrogen gas. The substrate was placed on the substrate holder which the target-to-substrate distance is fixed with 7cm and was uniformly rotated in the stainless steel vacuum chamber. The mechanical shutter was used for isolating the target material. The vacuum chamber was pumped by a turbo molecular to $5 \times 10^{-6}$ Torr and the pure argon and oxygen gas mixture was introduced through mass flow controllers to achieve the required pressure. The temperature of the substrate was varied from room temperature to 500. Before deposition, pre-sputtering for 10mins was performed to remove any contaminants and other particles on the target materials for better reproducibility. The structural properties of the films were analyzed with an X-ray diffractometer (Shimadzu, Inc., XRD6000) using Cu-Kα radiation. The XRD patterns were collected over a range of 10-80. The electrical resistivity was measured by a standard four-point probe method (DFP-03, SES instruments). Atomic force microscope (Nanoscope IV) studies were conducted to analyze the morphology of the films. The optical transmittance was measured in the visible light wavelength (400-800nm) with Ocean Optics HR4000 spectrophotometer.

## RESULTS AND DISCUSSIONS

*Effect of the RF Power*

In order to investigate the effect of RF power on ITO films, a range of RF power from 30W to 70W were applied at 5mTorr for the working pressure, 20min, and without oxygen flow rate. Figure 1 shows the deposition rate with different values of the RF power. As the RF power is increased, the deposition rate is also increased. It is considered that the energetic particles accelerated in the electric field would have higher kinetic energy to generate more sputtered particles as the RF power is increased.

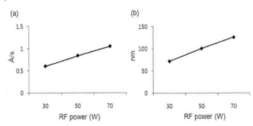

Figure 1. Deposition rate (a) and thickness (b) as a function of RF power

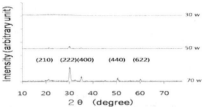

Figure 2. X-ray peaks of ITO films deposited at different values of RF power (W)

The X-ray diffraction patterns of the ITO films deposited at a various RF power are shown in Figure 2. It can be seen that the X-ray diffraction pattern of the ITO film deposited at the RF power of 30W indicates amorphous state. The ITO film deposited at 50W shows the small (222) peak of $In_2O_3$. There is an appearance of (400) peak of the ITO film when the RF power reached 70W. It also can be seen that the intensity of the peak becomes sharper. It is also known that the thickness of ITO film affects on X-ray spectra [10]. According to literature, the change in the orientation from (222) to (400) can be influenced by the deposition rate because the (222) oriented grains are less resistant against sputtering than the (400) oriented grains [10].

The transmittance of the ITO films as a function of the light wavelength is given in Figure 3(a) for ITO films deposited at various RF power values. It can be seen that as the RF power increases, the average transmittance is decreased. The average transmittance of ITO film deposited at 30W shows 70% of transmittance, but the average transmittance of ITO film deposited at 70W shows 58% of transmittance. It is understood by the effect of thickness since the optical scattering increases as the optical path which is the thickness of the ITO films is increased[11-13].

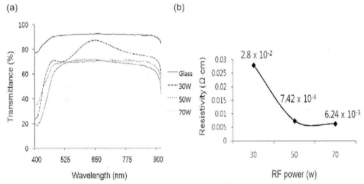

Figure 3.Optical transmittance (a) and electrical resistivity (b) of the ITO films deposited at different values of RF power

Figure 3(b) shows the electrical resistivity of the ITO films deposited at various RF power values. The electrical resistivity is decreased as the RF power is increased. The conductivity of the ITO film is fundamentally from its oxygen vacancies and Sn doping. It is reported that the deposited ITO films by sputtering without oxygen have lower the oxygen concentration than that of the target, since a part of the oxygen from the target material is not perfectly incorporated in the deposited films[14].

*The Effect of the Working Pressure*

Figure 4 shows the effect of working pressure on the deposition rate. The ITO films were deposited at 70W RF power, for 20min, and with various values of the working pressure, from 3mTorr to 7mTorr. The deposition rate is decreased as the working pressure is increased. It is believed that the higher working pressure induces more collisions between the sputtered particles and working gas. Thus, the mean free path decreases because the sputtered particles undergo more collisions with working gas and it results in losing more kinetic energy[15]. Therefore, the surface mobility is decreased and the deposition rate is also decreased.

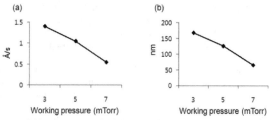

Figure 4. Deposition rate (a) and thickness (b) as a function of working pressure

The dependence of the structural properties of the ITO films on the deposition pressure is depicted in Figure 5. (222) and (400) peaks are observed for the ITO films deposited below 7mTorr of the working Ar pressure. However, the (400) peak is much smaller than the (222) peak for each sample. The intensity of (400) peak is decreased while the working pressure is increased. Finally the X-ray peak shows the amorphous state when the working pressure reaches a value of 7mTorr. It can be seen that the ITO films deposited below 7mTorr are polycrystalline. The fact of that the (400) peak is decreased as the working pressure is increased can be due to the thickness. The thickness of the ITO film deposited at 3mtorr is higher than that of 5mtorr. It has been reported that the thicker films tends to have higher (400) peak than thinner films[15]. In addition, it also can be attributed to the higher kinetic energy of the sputtered particles at a lower pressure. As mentioned before, the sputtered particles at a lower pressure tend to have higher kinetic energy to influence on the growth on the substrate[16].

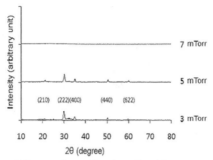

Figure 5. X-ray peak as a function of working pressure

Figure 6(a) shows the transmittance of the deposited ITO films as a function of the working pressure. It is found that the transmittance is increased as the working pressure is increased. This is attributed to the increase of deposition rate and thickness with a decrease of the working pressure. As previously reported, it is understood by the effect of thickness since the optical scattering increases as the optical path which is the thickness of the ITO films is increased[11-13]. From the decrease of the transmittance at the near infrared with increase of the working pressure, the carrier concentration is expected to be higher for the ITO film deposited at higher working pressure than that of lower pressure.

This is attributed that the deposition rate of the ITO film deposited at a lower pressure is higher than that of the ITO film grown at a higher pressure. It has been reported that the deposition rate affects on the defect of oxygen vacancy, giving two extra free electrons for the conduction[12, 13].

The electrical resistivity shown in Figure 6(b) is consistent with this expectation. As seen in Figure 6(b), the ITO film prepared at 3mtorr shows the smallest electrical resistivity, 2.3 x $10^{-3}$ $\Omega$cm. It is well known that electrical resistivity is inversely proportional to carrier concentration and mobility.

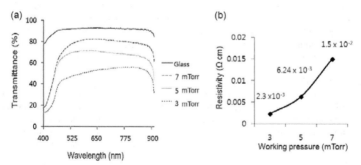

Figure 6. Optical transmittance (a) and electrical resistivity (b) of the ITO films deposited at different values of the working pressure.

*The Effect of Oxygen Partial Pressure*

In order to optimize the transmittance and the electrical resistivity of ITO films, different values of oxygen partial pressure were introduced to the chamber with the fixed amount of Ar, 3mTorr. The values of RF power and deposition time were constant with 70W and 20min, respectively. The deposition rate is not changed significantly and the thicknesses of the ITO films are maintained near 160nm range. The effect of oxygen partial pressure on the transmittance of deposited ITO films is given in Figure 7(a). It is shown that the transmittance is increased significantly as the oxygen partial ratio is increased to 2%. The increase of the oxygen partial ratio shows the increase of the transmittance. The highest transmittance is obtained at 2% of the oxygen partial ratio.

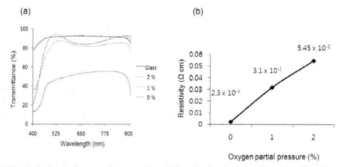

Figure 7. Optical transmittance (a) and electrical resistivity (b) of the ITO films deposited at different values of the oxygen partial pressure

Figure 7(b) shows the electrical resistivity as a function of the oxygen partial pressure. It can be seen that an increase of the electrical resistivity with increasing the oxygen partial pressure. The ITO film deposited at 1% of oxygen partial pressure shows $4.1 \times 10^{-2}$ $\Omega$cm value. The resistivity increased to $6.4 \times 10^{-2}$ $\Omega$cm as the oxygen partial pressure is increased to 2%. This is attributed to that the amount of oxygen vacancies of ITO film increases with increasing the oxygen partial pressure in the deposition process. It is knows that the high conductivity of the ITO films have been attributed to both doped tin and oxygen vacancies[2]. It is inferred that the ITO film deposited at a higher value of oxygen partial pressure has less amount of carrier concentration, leading higher electrical resistivity since the electrical resistivity is proportional to the carrier concentration. This is consistent with the data from Figure 7(b).

*The Effect of the Substrate Temperature*

To investigate the effect of the temperature of the substrate on the properties of ITO films deposited with RF reactive sputtering, the temperature of the substrate was varied from room temperature to 500 °C with an increment of 100 °C. During the deposition, the power, the working pressure, the oxygen partial pressure, and deposition time were maintained 70W, 3mTorr, 1%, and 10min respectively. All films have a value of near 75nm. Thus, the deposition rate is not changed significantly as an increase of the substrate temperature.

XRD patterns of ITO films deposited at different values of the substrate temperature are shown in Figure 8. It can be seen that the ITO film deposited at room temperature shows amorphous characteristic. As the temperature is increased, the XRD peaks showing the cubic bixbyite structure of $In_2O_3$ appear and become intense. It shows that the ITO films deposited above 100 °C have a preferred orientation of (222) plane. The further increase of the substrate temperature induces the XRD peak of (400) plane significant. The ITO film deposited at 400 °C shows increased (400) peak. It has been reported that the adatom mobility and kinetic energy of the sputtered adatoms are strongly related to the preferred orientation of ITO films. According to literature[17, 18], the change of preferred orientation from (200) to (400) occurs as the substrate temperature is increased.

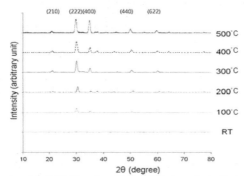

Figure 8. X-ray peaks of the ITO films deposited at different values of substrate temperature

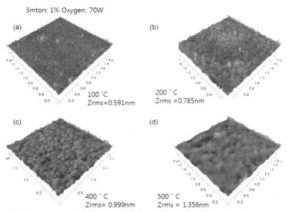

Figure 9. AFM images of the ITO films deposited at different values of the substrate temperatures

The AFM images, Figure 9, show that the grain size and roughness (Zrms) are increased as the increase of substrate temperature. This is related to the mobility of the sputtered ITO particles on the substrate. The mobility of the deposited ITO particles on the substrate is expected to increase with increasing of the substrate temperature. This mobility caused the adatoms to form bigger grains. Thus the roughness is also increased.

The transmittance of the ITO films prepared at various substrate temperatures is presented at Figure 10(a). All films show high transmittance around 80%. Figure 10(a) indicates that the transmittance of ITO films is decreased slightly in the visible region as the substrate temperature is increased. This can be attributed to the increased roughness with increasing the substrate temperature as seen in Figure 9. The rougher surface causes the more light to be scattered[19]. The carrier concentration can be inferred from the transmission value at the near the infrared region. It is known that the transition from high transmittance to high reflectance occurs at shorter wavelength as the carrier concentration is increased[20]. Thus, one can infer that the carrier concentration is increased as the substrate temperature is increased. In ITO films, there are two donors, giving free electron: oxygen vacancies and tin doping. Since the deposition was performed in the environment of continuous introducing of oxygen gas, it can be assumed that the change of the electron concentration from the oxygen vacancies would not be changed significantly. Therefore, the tin atoms dominate the carrier concentration. The high substrate temperature causes more tin atoms to diffuse along grain boundary, resulting in higher carrier concentration[21].

The electrical resistivity of the ITO films deposited at various substrate temperatures is indicated in Figure 10(b). It is shown that the electrical resistivity decrease as the substrate temperature is increased until 400°C and the electrical resistivity is increased as the substrate temperature is increased further. From Figure 10(a), it was shown that the carrier concentration increases with an increase of the substrate temperature. It is related to the fact of that the conductivity is also a dependent of the carrier mobility. Thus, it may be due to the decrease of the carrier mobility above 400°C.

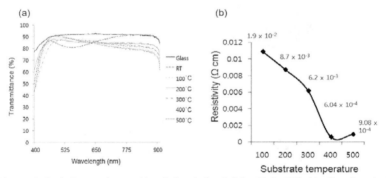

Figure 10. Optical transmittance (a) and electrical resistivity (b) for ITO thin films deposited at different values of the substrate temperature

CONCLUSION

ITO thin films were deposited on glass substrate by RF magnetron sputtering. The structural, electrical, and optical properties of the ITO films depended on the sputtering power, working pressure, oxygen partial pressure, and substrate temperature. The deposited ITO films showed polycrystalline and retained bixbyite structure of $In_2O_3$.

It was found that as the RF power increased, the transmittance decreased but resistivity increased. The crystallization was possible with an increased RF power. The X-ray peaks become more intense as the RF power is increased. The higher RF power causes the more sputtered particles and the higher mobility on the substrate, leading crystallization. Without introducing oxygen, ITO films showed poor transparency. The electrical resistivity was also related to the transmittance at near infrared region, indicating the lower value of the transmittance at the near infrared region means the higher value of the carrier concentration.

In investigation of the effect of pressure, the transmittance and electrical resistivity of the ITO films showed an increase with an increase of the working pressure. As the working pressure is increased, the sputtered particles would have more collisions with the working gas, losing the kinetic energy of the sputtered particles. It is considered that the collided sputtered particles would deposit on the chamber wall rather than on the substrate. It also reduced the mobility of the adatoms on the substrate and the degree of the crystallization. The X-ray peaks showed the amorphous characteristic of the ITO film deposited at 7mTorr. The transmittance and electrical resistivity of the ITO films were also a function of the working pressure. The higher transmittance value was obtained at the higher working pressure. The transmittance values at the near infrared regions expected that the higher value of the carrier concentration of the ITO films deposited at the lower value of the working pressure. The amount of the oxygen vacancies is expected to decrease as the working pressure is increased. This is due the decreased deposition rate with an increase of the working pressure. The lower deposition rate tends to possess less oxygen vacancies which donate maximally two electrons for conduction. It is known that optical scattering occurs at the center of the defects or impurities such as oxygen vacancies in ITO films. It was also shown that the electrical resistivity is increased as the working pressure is increased.

However, all ITO films deposited without oxygen showed poor transparency. The High transmittance around 80% in the visible region was obtained with introducing of oxygen mixed Ar gas. It is seen that the small amount of the oxygen partial pressure is able to degrade the conductivity of the ITO films. In this study, 1% of oxygen partial pressure was chosen as the optimized rate for both high transparency and high conduction.

In the study of the effect of the substrate temperature on the properties of the ITO films, it was shown that the substrate temperature affects on the crystallization of the deposited ITO films. The preferential orientation of (400) plane became significant when it was deposited above 400 °C. The optical transmittance showed a decreasing trend with an increase of the substrate temperature. The lowest electrical resistivity was obtained at 400 °C. It is believed that the increase of the electrical resistivity of the ITO film deposited at 500 °C is due to the decrease of the mobility.

Through this study, it was found that the optical, electrical and structural properties are strongly related to the sputtering parameters. We were able to confirm that all sputtering parameters are correlated to the properties of the deposited ITO films. The optimized condition for high transmittance and low electrical resistivity is that 70W of the RF power, 3mTorr of the working pressure, 1% of the oxygen partial pressure, and 400 °C of the substrate temperature.

With this condition, the deposited ITO film showed high value of transmittance, 80% and the comparable value of electrical resistivity, $6.04 \times 10^{-4}$ $\Omega$cm, having two dominant X-ray peaks of (222) and (400) planes.

ACKNOWLEDGEMENT
This work has been supported in part by *Texas STARS grant, Materials Research Graduate Fellowships* and *Robert E. Clarke Endowment Fund.*

REFERENCES
[1]W. Wu, B. Chiou and S. Hsieh, Effect of sputtering power on the structural and optical properties of RF magnetron sputtered ITO films, *Semicand. Scl. Technol.*, **9**, 1242-1249, (1994).

[2]H. Kim, C. Gilmore, A. Pique, J. Horwitz, H. Mattoussi, H. Murata, Z. Kafafi, and D. Chrisey, *J. Appl. Phys.*, **86**, 6451 (1999).

[3]J. Fan, F. Bachner, and G. Foley, Effect of Oxygen Partial Pressure During Deposition on Properties of r.f. Sputtered Sn-Doped In2O3 Films. *Applied Physics Letters.* **31(11)**, 773 - 775 (1977).

[4]J. Szczyrbowski, A. Dietrich, and H. Hoffmann, Optical and Electrical Properties of r.f. Sputtered Indium-Tin Oxide Films. *Phys. Stat. Sol.* **78**, 243 - 252 (1983).

[5]P. Nath and R. Bunshah, Preparation of In2O3 and Tin-Doped In2O3 Films by a Novel Activated Reactive Evaporation Technique. *Thin Solid Films.* **69**, 63-68 (1980).

[6]H. Haitjema, and J. Elich, Physical Properties of Pyrolitically Sprayed Tin-Doped Indium Oxide Coatings. *Thin Solid Films.* **205**, 93-100 (1993).

[7]S. Tang, J. Yao, J. Chen, and J. Luo, Preparation of indium tin oxide (ITO) with a single-phase structure. *Journal of Materials Processing Technology.* **137**, 82-85 (2003).

[8]U. Betz, M. Olsson, J. Marthy, M. Escola, and F. Atamny, *Surf. Coat. Technol.* **200**, 5751 (2006).

[9]N. Danson, I. Safi, G. Hall, and R. Howson, *Surf. Coat. Technol.* **99**, 147 (1998).

[10]B. Cullity, Elements of X-Ray Diffraction. *Addison-Wesley Publishing Company, Boston, Mass, USA, 2nd edition.* (1978)

[11]S. Jun, T. McKnight, M. Simpson, P. Rack, A statistical parameter study of indium tin oxide thin films deposited by radio-frequency sputtering. *Thin Solid Films.* **476 (1)**, 59-64 (2005).

[12]C. Guillén, and J. Herrero, Polycrystalline growth and recrystallization processes in sputtered ITO thin films. *Thin Solid Films.* **510 (1-2)**, 260-264 (2006)

[13]H. Kim, J. Horwitz, and G. Kushto G.) Effect of film thickness on the properties of indium tin oxide thin films. *Journal of Applied Physics.* **88 (10)**, 6021-6025 (2000).

[14]J. Vetrone, and Y. Chung, Organo-metallic chemical vapor deposition of tin oxide single crystal and polycrystalline films. *Journal of Vacuum Science and Technology A.* **9 (6)**, 3041-3047 (1991).

[15]L. Meng, M, Andritschky, M. dos Santos, The effect of substrate temperature on the properties of d.c. reactive magnetron sputtered titanium oxide films. *Thin Solid Films.* **223 (2)**, 242-247 (1999).

[16]H. Kersten, G. Kroesen, R. Hippler, On the energy influx to the substrate during sputter

deposition of thin aluminium films. *Thin Solid Films.* **332 (1-2)**, 282-289 (1998).

[17]T. Vink, W. Walrave, J. Daams, P. Baarslag, J. van den Meerakker, On the homogeneity of sputter-deposited ITO films Part I. Stress and microstructure. *Thin Solid Films.* **266 (2)**, 145-151 (1995).

[18]S. Uthanna, P. Reddy, B. Naidu, P. Reddy, Physical investigations of DC magnetron sputtered indium tin oxide films. *Vacuum.* **47 (1)**, 91-93 (1996).

[19]A. Doron, E. Katz, I. Willner, Organization of Au Colloids as Monolayer Films onto ITO Glass Surfaces: Application of the Metal Colloid Films as Base Interfaces To Construct Redox-Active Monolayers. *ACS Publications.* **11 (4)**, 1313-1317 (1995).

[20]M. Liua, and S. Chou, High-modulation-depth and short-cavity-length silicon Fabry–Perot modulator with two grating Bragg reflectors. *Appl. Phys. Lett.*, **68**, 170 (1996).

[21]R. Tahar, T., Ban, T., Ohya, Y. Takahashi. Y. Tin doped indium oxide thin films: Electrical properties. *J. Appl. Phys.*, **83**, 2631 (1998).

SIMULATION OF ENHANCED OPTICAL TRANSMISSION IN PIEZOELCTRIC
MATERIALS

Robert McIntosh, Amar S. Bhalla, and Ruyan Guo*
Multifunctional Electronic Materials and Devices Research Lab
Department of Electrical and Computer Engineering
The University of Texas at San Antonio, One UTSA Circle, San Antonio, TX 78249, USA

ABSTRACT
    A preliminary study is conducted on the impact of large ferroelectric remnant polarization on electromagnetic wave propagation. Both experimental and computational evaluations show enhanced propagation resulting from a structure with periodic bound charge distribution. A finite element model has been constructed for evaluating the effectiveness of this proposed structure. The enhancement in transmission is directly related to the strain-coupled local polarization. At piezoelectric resonance oscillating dipoles or local polarizations become periodic in the material and have the greatest impact on transmission. Preliminary results suggest that the induced charge distribution by a piezoelectric material at certain resonant frequencies is effective for altering the transmission of a propagating wave. The behavior of both piezoelectric defined and engineered periodic structures are reported.

INTRODUCTION
    The interaction between light and piezoelectric materials is very important for optical communications. All piezoelectric materials are also electro optic and photoelastic. Photoelasicity can be expressed in terms of piezo optic or elasto optic effects. It has long been know that the electro optic output is strongly influenced by low frequency (KHz) vibrations due to natural piezoelectric resonate frequencies[6]. Typically electro optic modulators are not operated in the region of resonance vibration due to the non-monotonic response however it has been shown that operation at these frequencies and at harmonics can actually be advantageous to improve electro optic response[2]. The response of the electro optical signal tends to follow the slope of the magnitude of the electrical admittance, this is fairly unsurprising as the largest mechanical vibrations occur at resonance (typically between the resonate and antiresonate frequencies) and in turn cause a large change in refractive index. Further results by Guo et al have shown that certain resonate modes with vibration mostly in the directions transverse to optical propagation have a larger influence on response because they are parallel to the optical polarization[1]. Given that the spurious response of these crystals little attention has been made to the detailed study of their response in conjunction with the optical field around resonance, leaving it an open and promising area of research for further optimization of electro optic devices. To study this interaction a model has been developed using Comsol Multiphysics for Finite Element Analysis (FEA). Due to this many physics solver the complex nature of this problem can be examined in detail incorporation mechanical deformation, photoclasticity, and electro optics.

THE SAMPLE
    The material used was $(1-x)Pb(Mg_{1/3}Nb_{2/3})O_3-(x)PbTiO_3$ where $x = 0.045$ (PMN-30%PT) which is a single crystal poled along <001>. At room temperature this composition is

rhombohedral, however after poling it is known to be pseudo-tetragonal (4mm symmetry)[3]. The sample is electrode on its (001) faces (Figure 1).

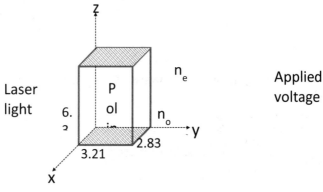

**Dimensions in mm**

Figure 1, PMN-30%PT sample, poled in [001] direction. Electroded on top and bottom surfaces formed by the x and y dimensions. The electrode area = 3.21 mm x 2.83 mm=9.0843 x$10^{-6}$ m$^2$, thickness=6.3x$10^{-3}$ m. The incident optical signal propagates along the y axis.

THE PIEZOELCTRIC MODEL

The piezoelectric model is used to identify the primary resonate modes of the sample and eventually when applied to the optical model to calculate the refractive index changes. Equation (1) is the partial differential equation used to describe the relationship between the density ($\rho$), angular frequency ($\omega$), mechanical displacement (u) and stress (X) in the piezoelectric domain (for a time harmonic solver). Additionally in the domain, the divergence of the electric displacement (D) field is equal to the charge density ($\rho_v$). The electric field is equal to the gradient of the negated voltage and the strain ($x$) is related to the mechanical displacement.

$$-\rho\omega^2 u - \nabla \cdot X = F_v e^{j\varphi} \tag{1}$$

$$\nabla \cdot D = \rho_v \tag{2}$$

$$E = -\nabla V \tag{3}$$

$$x = \frac{1}{2}\left[(\nabla u)^T + \nabla u\right] \tag{4}$$

This simulation uses the converse piezoelectric effect, meaning an applied electric field causes a change in the strain. In this case in equations (5) and (6) the Stress-charge form is used. Respectively the terms $X$, $C_E$, $x$, $e$, $\varepsilon_0$, $\varepsilon_{rs}$, $D_r$ are the stress, elasticity matrix, strain, coupling matrix, vacuum permittivity, relative permittivity, and remanent displacement field. The $x_i$ term is the initial stress in the material and is assume to be zero.

Stress-Charge form:

$$X = c_E(x - x_i) - e^T E + X \tag{5}$$

$$D = e(x - x_i) - \varepsilon_0 \varepsilon_{rS} E + D_r \tag{6}$$

The pertinent material parameters for a piezoelectric material with 4mm symmetry are shown in Equation (7). The symmetry is applied to the elasticity matrix ($c_{ij}$), the piezoelectric coupling matrix ($e_{ij}$) and the permittivity matrix ($\varepsilon_{ij}$). Table I Shows the material parameters used in the piezoelectric simulation. Literature values are from Zhang $el\ al.$[7]

$$c_{ij} = \begin{bmatrix} c_{11} & c_{12} & c_{13} & 0 & 0 & 0 \\ c_{12} & c_{11} & c_{13} & 0 & 0 & 0 \\ c_{13} & c_{13} & c_{33} & 0 & 0 & 0 \\ 0 & 0 & 0 & c_{44} & 0 & 0 \\ 0 & 0 & 0 & 0 & c_{44} & 0 \\ 0 & 0 & 0 & 0 & 0 & c_{66} \end{bmatrix} \quad e_{ij} = \begin{bmatrix} 0 & 0 & 0 & 0 & e_{15} & 0 \\ 0 & 0 & 0 & e_{15} & 0 & 0 \\ e_{31} & e_{31} & e_{33} & 0 & 0 & 0 \end{bmatrix} \quad \varepsilon_{ij} = \begin{bmatrix} \varepsilon_{11}' - j\varepsilon^{\cdot} & 0 & 0 \\ 0 & \varepsilon_{11}' - j\varepsilon^{\cdot} & 0 \\ 0 & 0 & \varepsilon_{33}' - j\varepsilon^{\cdot} \end{bmatrix} \tag{7}$$

**Table I,** *Reported values for the Elasticisy, coupling, real permititiy, and density are from Zhang *et al.* The imagonary permititiy is a measured values for this sample.

| Elasticity (@ constant E-filed) [N/m²]* | Coupling matrix [C/m²]* | Permittivity (@ constant temperature)* | Density [kg/m³]* |
|---|---|---|---|
| $c_{11}$=11.7x10¹⁰ | $e_{15}$=13.6 | $\varepsilon_{11}'$=3307 | 8038.4 |
| $c_{12}$=10.3x10¹⁰ | $e_{31}$=-2.4 | $\varepsilon_{33}'$=1242 | |
| $c_{13}$=10.1x10¹⁰ | $e_{33}$=27.1 | | |
| $c_{33}$=10.8x10¹⁰ | | $\varepsilon''=1$ | |
| $c_{44}$=7.1x10¹⁰ | | | |
| $c_{66}$=6.6x10¹⁰ | | | |

A 0.5 volt AC source is used to excite the sample and the bottom domain is grounded. The remaining four boundaries are all under zero charge condition. The crystal is completely unconstrained on all boundaries, edges and points. In order to calculate the response of the electrical conductivity the real ($I_r$) and imaginary ($I_i$) parts of the current were integrated over the surface of the top boundary (the boundary of the electrode). Using these values and combing with the applied AC voltage ($V_{app}$=0.5) gives the conductance (G) in equation (8) and susceptance in (B) equation (9). Then the Magnitude of the admittance can easily be calculated in equation (10) and the phase angle in equation (11)

$$G = I_r / V_{app} \tag{8}$$
$$B = I_i / V_{app} \tag{9}$$
$$|Y| = \sqrt{G^2 + B^2} \tag{10}$$
$$Phase = Tan^{-1}(B/G) \tag{11}$$

The results of the admittance calculation are shown in
Figure **2**, resonance frequencies are indicated by peaks in the |Y| and minimums indicate anti resonate frequencies. Additionally the sample was also tested experimentally and a comparison

is show in **Figure 3**. The response does not line up well however the amplitudes are on a comparable scale and tests/simulations on similar compositions of higher dimensional aspect ratio have shown extremely close agreement.

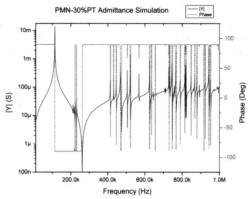

Figure 2, Simulation of Admittance magnitude |Y| in Siemens and the phase angle in degrees over frequency range encompassing the lowest piezoelectric vibrational frequencies.

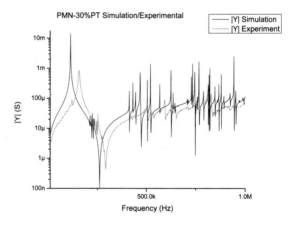

Figure 3, Comparison of the Simulation and Experimental results of the admittance frequency spectrum of PMN-30%PT.

THE OPTICAL MODEL

The optical model consists of two parts one being a three dimensional model identical to the piezoelectric model described above and the other being a two dimensional model used to compute the transmission parameters of the optical transmission through the center of the piezoelectric crystal (See Figure **4**). This model uses a slightly different sample, with electrode area of 2.26 mm x 2.52 mm=5.6952 x$10^{-6}$ m$^2$, and thickness=2.05x$10^{-3}$ m. This sample is PZN-4.5%PT and possesses the same pseudo tetragonal 4mm symmetry as the previous sample but with higher opacity lending itself well to optical testing.

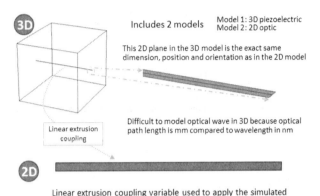

**Figure 4.** The Optical simulation model includes two sub models. Model 1 is formed in 3 space dimensions using the same parameters as described in the piezoelectric modeling section. Model 2 uses only 2 space dimensions to allow for more efficient examination of the optical transmission. Linear extrusion coupling variable are used to map the refractive index changes from the 3D to the 2D model.

The Optical model is rectangular with the longer dimensions being the direction of wave propagation (y-axis). The side boundaries are perfect magnetic conductors allowing plain wave propagation. The end boundaries are ports, one port is the output and the other is used to launch a 5mW wave polarized in the x and z directions (transverse to propagation). The Propagation constant is β=(2*π)/(3x$10^8$[m/s]/2.857x$10^{14}$ [Hz]), where 2.857x$10^{14}$ Hz is the frequency of a 1050 nm wave in free space. The partial differential equation applied to the domain is:

$$\nabla \times \mu_r^{-1}(\nabla \times E) - k_0^2 \left( \varepsilon_r - \frac{j\sigma}{\omega\varepsilon_0} \right) E = 0 \qquad (12)$$

Where $\mu_r$, $\varepsilon_r$, $\varepsilon_0$, $k_0$, $\sigma$, $\omega$, $E$ are respectively the vacuum permeability, relative permittivity, vacuum permittivity, free space wave number, conductivity, angular frequency and the electric field. The refractive index changes due to the applied electric field and the shape changes of the piezoelectric sample are modeled by way of the electro optic and elastooptic effects respectively. The electro optic effect is the change of the refractive index due to an applied electric field $(E)$ and at constant strain (the clamped condition) is represented by $r^x$ the elastooptic effect is the change in refractive index due to a strain $(x)$ and is represented by $p$. Combining these effects and defining in tensor form the impermeability tensor $B_{ij}$ is:

$$\Delta B_{ij} = r_{ijk}^x E_k + p_{ijkl} x_{kl} \tag{13}$$

$\Delta B_{ij}$ can be defined in terms of the refractive index with zero applied field and zero stress $(n_{ij})$ and the new refractive index $n_{ij}'$

$$\Delta B_{ij} = \Delta \left( \frac{1}{n_{ij}^2} \right) = \left( \frac{1}{n_{ij}'} \right)^2 - \left( \frac{1}{n_{ij}} \right)^2 \tag{14}$$

By combining equations (13) and (14), converting to matrix form and applying 4mm symmetry the matrix for the six independent refractive indices is:

$$
\begin{bmatrix}
\left(\frac{1}{n_{11}'}\right)^2 \\
\left(\frac{1}{n_{22}'}\right)^2 \\
\left(\frac{1}{n_{33}'}\right)^2 \\
\left(\frac{1}{n_{23}'}\right)^2 \\
\left(\frac{1}{n_{31}'}\right)^2 \\
\left(\frac{1}{n_{12}'}\right)^2
\end{bmatrix}
=
\begin{bmatrix}
\left(\frac{1}{n_{11}}\right)^2 \\
\left(\frac{1}{n_{22}}\right)^2 \\
\left(\frac{1}{n_{33}}\right)^2 \\
\left(\frac{1}{n_{23}}\right)^2 \\
\left(\frac{1}{n_{31}}\right)^2 \\
\left(\frac{1}{n_{12}}\right)^2
\end{bmatrix}
+
\begin{bmatrix}
0 & 0 & r_{13} \\
0 & 0 & r_{13} \\
0 & 0 & r_{33} \\
0 & r_{51} & 0 \\
r_{51} & 0 & 0 \\
0 & 0 & 0
\end{bmatrix}
\begin{bmatrix}
E_1 \\
E_2 \\
E_3
\end{bmatrix}
+
\begin{bmatrix}
p_{11} & p_{12} & p_{13} & 0 & 0 & 0 \\
p_{12} & p_{11} & p_{13} & 0 & 0 & 0 \\
p_{31} & p_{31} & p_{33} & 0 & 0 & 0 \\
0 & 0 & 0 & p_{44} & 0 & 0 \\
0 & 0 & 0 & 0 & p_{44} & 0 \\
0 & 0 & 0 & 0 & 0 & p_{66}
\end{bmatrix}
\begin{bmatrix}
x_{11} \\
x_{22} \\
x_{33} \\
x_{23} \\
x_{31} \\
x_{12}
\end{bmatrix}
\tag{15}
$$

Simplifying gives the following equation which is applied to the Comsol model to fully describe the coupling of the refractive index values between the piezoelectric to the optical models.

$$n'_{11} = 1 \Big/ \sqrt{\frac{1}{n_{11}^2} + r_{13}E_3 + p_{11}x_{11} + p_{12}x_{22} + p_{13}x_{33}}$$

$$n'_{22} = 1 \Big/ \sqrt{\frac{1}{n_{22}^2} + r_{13}E_3 + p_{12}x_{11} + p_{11}x_{22} + p_{13}x_{33}}$$

$$n'_{33} = 1 \Big/ \sqrt{\frac{1}{n_{33}^2} + r_{33}E_3 + p_{31}x_{11} + p_{31}x_{22} + p_{33}x_{33}} \qquad (16)$$

$$n'_{23} = 1 \Big/ \sqrt{r_{51}E_2 + p_{44}x_{23}}$$

$$n'_{31} = 1 \Big/ \sqrt{r_{51}E_1 + p_{44}x_{31}}$$

$$n'_{12} = 1 \Big/ \sqrt{p_{66}x_{12}}$$

The elasto optic, electro optic and refractive indices for the material were taken from the literature and are listed in Table II. The refractive indices and elctro optic coefficients are actually for a similar composition (PZN-12%PT). The elasto optic coefficients are for PZN-PT and where actually converted from the piezo optic coefficients in the literature with the equation $p_{mn} = \pi_{mp} c_{pn}$.

**Table II**, Optical parameters for PZN-PT. Elasto optic coefficients for PZN-4.5%PT from Lu et al.[4] Refractive indies and electro optic coefficients for PZN-12%PT from Lu et al.[5]

| Refractive indices | | Elasto optic coefficients | |
|---|---|---|---|
| ne | 2.57 | $p_{11}$ | 0.0888 |
| no | 2.46 | $p_{12}$ | 0.0816 |
| | | $p_{13}$ | 0.0808 |
| Electro optic coefficients | | $p_{31}$ | -1.032 |
| $r_{13}$ | 7 pm/V | $P_{33}$ | 1.911 |
| $r_{33}$ | 134 pm/V | $p_{44}$ | 0 |
| $r_{51}$ | 462 pm/V | $p_{66}$ | 0 |

The results of the optical field simulation can be seen in **Figure 5** below. The plane wave propagates along the entire length of the sample domain. The launched wave is 1050 nm in free space but for a refractive index of about 2.5 the wave is 420 nm in the material – this confirms that the meshing on the domain is dense enough to represent the wave.

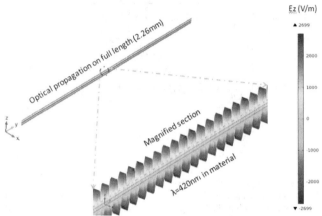

Figure 5, The Optical plane wave with propagation in the y direction and plot of the electric field ($E_z$) in the transverse z direction.

In **Figure 6** the transmission of the optical signal using the transmission parameter $S_{21}$ is shown with the admittance simulation. The figure clearly shows an increase the transmission near the resonant frequencies, typically the transmission follows the slope of the admittance magnitude and is quite consistent with typical experimental results for a sample near the resonant frequencies[1].

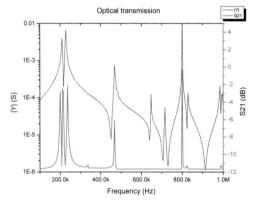

Figure 6, Frequency response of the optical response over range of low frequency resonate modes showing strong correlation with the slope of the electrical admittance.

At resonance the current field inside the material becomes periodic and leads to an induced polarization change inside the sample at integer multiples of the fundamental resonate modes. As

proposed by Johnson *et al*[2] it is this periodically changing field that has a large influence on the transmission and our model of the combined influence of the electro optic and elasto optic effects. As illustrated in
Figure 7 the current under certain resonant modes (in this case at 215 KHz) promotes a refractive index gradient that is sufficient to cause a large increase in transmission magnitude. Both acoustic and optical phonons can exist in the crystal. Polarization is present in the transverse optical phonons (TO) which couples with the displacive current. When at resonance the TO modes become periodic and result in a complete circuit of current flow in the local field.

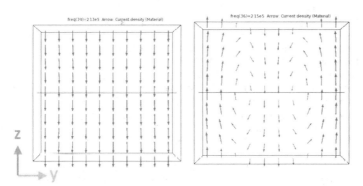

Figure 7, Current density distribution in the sample; off piezoelectric resonance at 213 KHz (Left) and near piezoelectric resonance at 215 KHz (Right). At certain resonance modes the current becomes periodic and leads to an induced changing charge distribution. At resonance more of the current is copropagating with the optical wave and enhancement is observed. The y-axis is the direction of propagation and z-axis the wave polarization direction.

CONCLUSION
It has been shown that under certain conditions of piezoelectric resonance the optical response of the material can be tuned. Experimental results on electro optic crystals show high correlation between crystal vibrational modes and electro optical transmission signal. A Comsol model was developed to describe the low frequency crystal admittance response, the combined effects of electro optic and elastooptics and transmission of an optical wavelength plane wave. The Comsol model is in good agreement with experimental results for both vibration modes and optical response. Previously proposed models used to explain the influence of bound charge distribution on wave transmission have been given further credence by multiphysics modeling of the electric displacement current influence on the propagating optical wave.

ACKNOWLEDGEMENT
This work has been partially supported by Office of Naval Research under grant number N00014-08-1-0854 and by NSF under grant# 1002380.

REFERENCES

[1]R. Guo, H. Liu, G. Reyes, W. Jamieson, and A. Bhalla, "Piezoelectric resonance enhanced electrooptic transmission in PZN-8PT single crystal," pp. 705616. **Vol. 7056** Edited by Y. Shizhuo and Ruyan Guo..

[2]S. Johnson, K. Reichard, and R. Guo, "Dynamic linear electrooptic property influnced by piezoelectric resonance in PMN-PT crystals," pp. 277-87 in 106th Annual Meeting of the American Ceramic Society, April 18, 2004 - April 21, 2004. **Vol. 167,** *Ceramic Transactions*.

[3]P. D. Lopath, P. Seung-Eek, K. K. Shung, and T. R. Shrout, "Single crystal $Pb(Zn_{1/3}Nb_{2/3})O_3/PbTiO_3$ (PZN/PT) in medical ultrasonic transducers," pp. 1643-46 vol.2 in Ultrasonics Symposium, 1997. Proceedings., 1997 IEEE. **Vol. 2.**

[4]Y. Lu, Z. Y. Cheng, Y. Barad, and Q. M. Zhang, "Photoelastic effects in tetragonal $Pb(Zn1/3Nb2/3)O3-PbTiO3$ single crystals near the morphotropic phase boundary," *Journal of Applied Physics,* **89**[Copyright 2001, IEE] 5075-8 (2001).

[5]Y. Lu, C. Zhong-Yang, P. Seung-Eek, L. Shi-Fang, and Z. Qiming, "Linear electro-optic effect of $0.88Pb(Zn1/3Nb2/3)O3-0.12PbTiO3$ single crystal," *Japanese Journal of Applied Physics, Part 1 (Regular Papers, Short Notes & Review Papers),* **39**[Copyright 2000, IEE] 141-5 (2000).

[6]Y. Pisarevski and G. Tregubov, "The Electro-Optical Properties of NH4H2PO4, KH2PO4, and NC(CH2)6 Crystals in UHF Fields," *Soviet Physics-Solid State,* **7**[2] (1965).

[7]R. Zhang, W. Jiang, B. Jiang, and W. Cao, "Elastic, dielectric and piezoelectric coefficients domain engineered 0.70Pb(Mg1/3Nb2/3)O3-0.30PbTiO3 single crystal," *AIP Conference Proceedings*[Copyright 2003, IEE] 188-97 (2002).

# EVOLUTION OF MICROSTRUCTURE DUE TO ADDITIVES AND PROCESSING

N. B. Singh, A. Berghmans, D. Knuteson, J. Talvacchio, D. Kahler, M. House, B. Schreib, B. Wagner and M. King

Northrop Grumman Corporation
ATL-3B10, 1212 Winterson Road
Linthicum, MD 21090
USA

ABSTRACT
    We have performed synthesis and growth of processing of $CaCu_3Ti_4O_{12}$ (CCTO) to achieve large size homogeneous grains in different thermal conditions. A temperature range of 900 to 1200C was used for processing. We studied the effect of excess Calcium oxide and addition of PbO on the morphologies of the grains. We observed that CaO rich CCTO remained non faceted similar to that of CCTO. While as addition of lead oxide generated very nice facetted morphologies. We observed that stoichiometric bulk CCTO showed dielectric constant in the range of 100,000 and loss tan delta in the range of 0.01 in stoichiometric CCTO materials. We observed that processing conditions and grain size had significant effect on the values of dielectric constant and loss tan delta.

## 1. INTRODUCTION

    A huge number of papers [1-4] have been published in literature on the development of $CaCu_3Ti_4O_{12}$ (CCTO) material for capacitors. Some of these papers have focused on single crystal material and a large number of materials are devoted to powder processed bulk and thin films. The limitation of today's CCTO is the leakage current. CCTO will be the first material to solve the problem for a material with permittivity of $10^5$ where the trade space for material uniformity, charge compensation, and design of composite structures is more constrained. The status of leakage current measurements in CCTO, expressed as a conductivity or resistivity, is in the range of $10^7$ $\Omega$-cm. A range of the resistivity values have been reported based on the processing conditions of the materials. All these values are reasonably lower than an ideal value needed for practical applications. The main objective for an R×C hold time requires a resistivity of about $10^{11}$ $\Omega$-cm. Several mechanisms have been put forward to explain the large dielectric constant and loss tan delta. Some of these mechanisms are based on the grains and single crystallinity of the processed CCTO. The Grain Boundary model of extremely high permittivity in CCTO was invoked in the literature cautiously at first, as a way to explain the indisputable presence of two simultaneous polarization mechanisms at work, with the larger component becoming inactive at high frequency (>10 MHz at room temperature) or low temperature (>10 kHz at 100 Kelvin). We have performed detailed studies on the growth of large grains of stoichiometric CCTO and non stoichiometric grains. During this processing of material we observed transition of non facetted to facetted grain morphologies for CCTO. In this paper we report preliminary results of these experiments.

## 2. EXPERIMENTAL METHODS

### 2.1 Materials Synthesis

We used traditional methods of preparing CCTO stoichiometric compound. In most of the cases, $CaCO_3$, $CuO$ and $TiO_2$ were used as the parent compounds for preparing the mixture. Powder was pressed to make pellets of cm sizes for processing. We used a temperature range of 1000 to 1200C for grain growth. In many experiments we used the temperature range of 1150+C to produce liquid phases for liquid Enhanced Ostwald ripening method. This process enables coarsening and grain growth by faster diffusion and mass transfer to achieve large materials in short time compared to ceramic processing.

For understanding the effect of liquid enhanced grain growth, we performed two sets of experiments by addition of PbO and CaO. The processing time and temperature range were kept identical. The morphology of material was studied by optical microscope and scanning electron microscope.

### 2.2 Fabrication Resistivity/Leakage Improvements

We have developed parameters for the cutting, polishing, electrode bonding and other fabrication of CCTO based capacitors. We were able to achieve good quality polished surfaces in automated polishers. We used several solvents as lubricants for achieving good quality polishing. We observed that using high viscosity solvents such as glycerol one can improve surfaces even better. The gold electrodes were placed using physical vapor deposition method in DENTON commercial evaporator. Material does not react with gold and long term stability is extremely good. We observed that using high viscosity solvents such as glycerol one can improve surfaces even better. Resistivity and dielectric constants were measured for pure, doped as well as samples prepared by liquid enhanced ripening method.

### 2.3 Dielectric Constant and Capacitance and resistance measurement techniques for bulk CCTO

Parallel-plate capacitors were formed on pressed and sintered pellets with thicknesses from 2 mm to 30 μm (thinned and polished). Typically, the entire back side of a sample was metalized whereas multiple small electrode pads were deposited on the front side for contacts to be made in a probe station. Electrodes were made with Ti/Au thin film bilayers or with silver paint.

Capacitance of bulk ceramic samples was inferred from measurements of complex impedance as a function of frequency from 20 Hz to 1 MHz with an Agilent/HP 4284A LCR meter. Resistivity was inferred from two measurements: (a) using a parallel-resistance model of complex-impedance measurements and (b) measuring current from a dc voltage bias. The latter measurement gave resistivity values that were larger by a factor of 1 to 2 times higher. Measurements were made as a function of temperature to 300°C and electric field to 20 V/μm.

## 3. RESULTS AND DISCUSSION

Since CCTO is a line compound with very little or no tolerance for deviations from the stoichiometry, we used stoichiometric 1:3:4 for Ca: Cu: Ti composition for all studies as base material. We used micron- and nano size particles as source materials for CaO, CuO and $TiO_2$. In the cases of carbonates, we used the calcinized powder for preparing CCTO. After mixing the powder we used a pressure range of …. for the cm size pellets. The furnace temperature was raised in steps. Better mixing and fewer voids were observed for materials compared to the materials prepared by fast heating process. Figure 1 shows a typical pellet of 12 mm size. When we used a traditional coarsening for 30-48 hours, the material produced large number of voids. Figure 2 shows a typical morphology. In spite of larger time of heat treatment, large number of holes remained in the material (Figure 3). CCTO prepared by a liquid enhanced ripening when temperature was raised above the melting point of copper oxide for 48+ hours of heat treatment. We observed rapid grain growth and sharp grain boundaries were observed compared to that in shown Figure 2. As the grains grew larger, smaller holes due to copper rich liquids were trapped. In some cases copper-rich small crystallites grew and it took long time for these crystallites to merge into larger grains. Figure 4 shows morphology of smaller elongated grains which appeared at the junction of the larger grains as well as on the top of smaller grains. Almost all the smaller grains are cylindrical and elongated with varying aspect ratio. It looks very long time for these grains to dissolve into larger grains.

**Figure 1.** Size and morphology of the CCTO material prepared for grain growth.

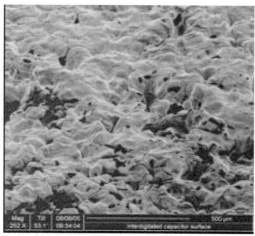

**Figure 2.** CCTO prepared by a traditional coarsening method at 1050C for 30+ hours of heat treatment

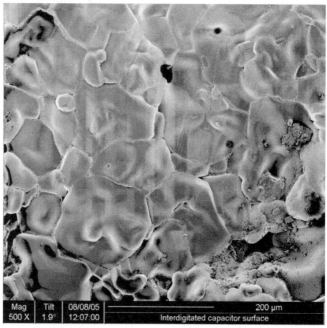

**Figure 3.** CCTO prepared by a liquid enhanced ripening when temperature was raised above the melting point of copper oxide for 48+ hours of heat treatment

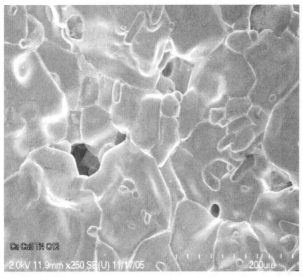

**Figure 4.** Morphology of smaller elongated grains appeared at the junction of the larger grains as well as on the top of smaller grains.

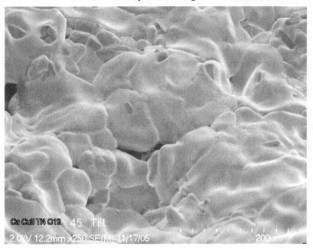

**Figure 5.** Morphology of tilted sample eat 45 degree, it was clear that grains grew on the top of each other to make larger grains and smaller grains in some cases separated and dissolved into a larger neighboring grains

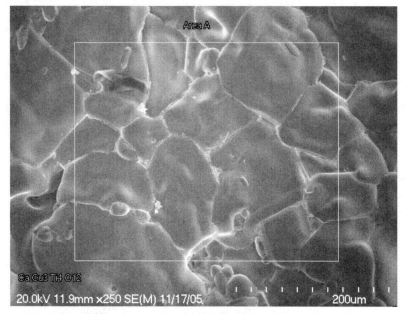

**Figure 6.** Morphology of the material processed at high temperature with no voids and holes.

**Figure 5** shows morphology of tilted sample at 45 degree. This figure shows that grains grew on the top of each other to make larger grains and smaller grains in some cases separated and dissolved into the larger neighboring grains. First time we observed the grains with sharper edges. In some places these grains appeared like particles separated and in some cases attached to larger grains. This picture shows thickening of grains and boundaries were not very sharp. They appear very much tilted and attached with some portion of the grains. Figure 6 shows morphology of the material processed at high temperature with no voids and holes. This show is clearly non facetted grains and boundaries. Smaller grains are still present on the boundaries. It is clear that a higher temperature and longer time is required to completely dissolve these grains located at the boundaries. Figure 7 shows x-ray diffraction pattern for a typical CCTO sample. It is clear that (004) peak is dominating. When we measured the composition of the material in the area A of the Figure 6, we did not find any measurable deviation from the stoichiometric composition.

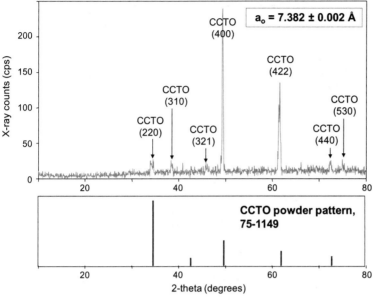

**Figure 7.** XRD pattern for a typical CCTO sample is shown in the Figure

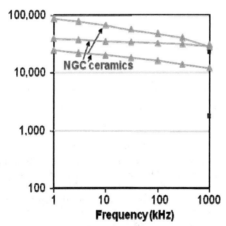

**Figure 8.** Dielectric values of three CCTO samples processed in three conditions. The dielectric constant of 100,000 was observed for the sample shown in Figure 6.

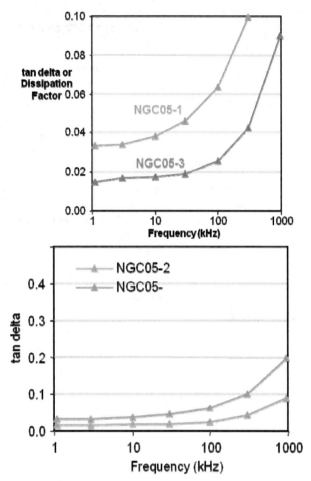

**Figure 9.** Tan delta or dissipation factor for two CCTO samples designated as NGC05-1 and NGC05-3.

First time ever in large size (cm+) processed sample a dielectric constant of 100,000 was observed as shown in Figure 8. Subramanian et al had indicated values in that range for a smaller CCTO single crystal sample. For CCTO, our experiments have greatly clarified the mechanism for its extremely high permittivity. However, the imaginary part of the dielectric constant, characterized by its ratio to the real part as the loss tangent or tan , is not well explained even for conventional dielectrics. Therefore, we will follow an empirical approach in tracking the loss tangent of CCTO and comparing it with the project goal of <0.001 at 1 kHz. Figure 9 shows the measured dissipation factor (ratio of the imaginary to real impedance) for a pair of CCTO ceramic samples designated as NGC05-1 and NGC05-3. The dissipation factor is often equated to tan delta, but high leakage current or series resistance due to the electrodes can make the dissipation much larger than tan delta. As the microstructures improved, loss tan delta improved and was observed well below 0.02. Figure 6 shows large grains in range of 100 μm free from bubbles and voids. Significant improvements were achieved in dielectric constant and dissipation factors of CCTO samples. Results of increased dielectric constant and decrease dissipation factors are shown in Figure 9 due to changes in process. A detailed paper on the measurements of dielectric constant and loss tan delta will be published in a future article. Very limited preliminary data are available. We collected complex impedance measurements as a function of temperature to >200°C. These measurements provide feedback on oxygenation, doping, and materials homogeneity. Some limited experiments on oxygenation and doping of CCTO showed increase in its resistivity, particularly in high electric fields >10 V/μm, to >$10^9$ Ω-cm.

We studied the morphologies by performing two separate experiments. In the first experiments we increased the concentration of calcium and decreased the concentration of copper. The calcium was increased from 1 to 1.04 and copper was decreased from 3 to 2.25. We used the identical processing to that of material shown in Figure 6. With increase of calcium we observed (Figure 10) that nature of grains completely changed. The grains were very irregular and we did not observe any sign of faceting. It appeared that due to lower concentration of copper, grain growth was slower and particles in the form of traps were observed. The growth of grains was very inhomogeneous. There were areas where we observed elongated grains of different sizes

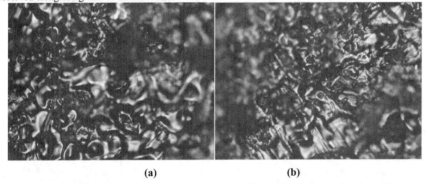

(a)                                    (b)

**Figure 10.** Microstructures (x100) of Ca-rich CCTO solid-liquid processed by Ostwald ripening method (non facetted morphology). The growth of grains was very inhomogeneous. There were areas where we observed elongated grains of different sizes.

We performed the growth of CCTO by adding very small amount of flux to increase the grain growth. For this purpose we added 6% PbO in the presynthesized CCTO compound for this study. After a treatment of 30 hours we observed growth of very facetted well shaped large grain growth at the expense of smaller grains. Figure 11 show morphology at different magnifications. Our observations at various parts of the sample showed the same morphology. In all cases grains were rectangular. It appears that some places growth occurred fast and grains grew in the range of 20 to 30 microns. The smaller grains remained in the one micron or smaller range. The mechanism appears similar to that shown in Figure 4-6 for pure CCTO. However in the case of PbO doped CCTO, grains are totally separated. There are sharp boundaries and we did not see large range of attached boundaries. Figure 12 shows morphology of a single grain. The surfaces look very rough and it appears as a disc before it forms rectangular crystallite. Figure 13 shows dissolution of a smaller grain into larger grains. However, it appears unlike non facetted grains very few grains merge into larger grains. A larger amount of PbO may enhance the formation of larger grains and may produce more homogeneous materials.

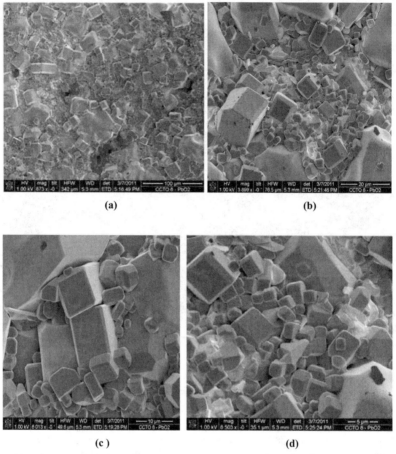

(a)

(b)

(c )

(d)

**Figure 11.** Microstructures of CCTO using Ostwald ripening in PbO solvent at several magnifications

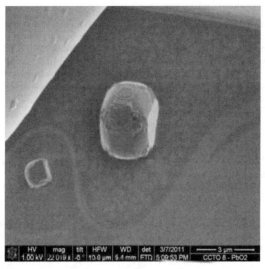

**Figure 12.** Morphology of a single grain dissolving into formation of a large grain

**Figure 13.** Formation of larger grains at the expense of smaller grains

4.  SUMMARY

Synthesis and growth of $CaCu_3Ti_4O_{12}$ (CCTO) was performed to achieve large size homogeneous grains in different thermal conditions. A temperature range of 900 to 1200C was used for processing and to achieve a dielectric constant in the range of 100,000 and loss tan delta in the range of 0.01 in stoichiometric CCTO materials... We studied the effect of excess Calcium oxide and addition of PbO on the morphologies of the grains. We observed that CaO rich CCTO remained non faceted similar to that of CCTO. While as addition of lead oxide generated very nice facetted morphologies. We observed that processing conditions and grain size had significant effect on the values of dielectric constant and loss tan delta.

5.  REFERENCES

1.  N. C. Homes, T. Vogt, S. M. Shapiro, S. Wakimoto, and A. P. Ramirez, "Optical response of high-dielectric-constant perovskite-related oxide," Science 293, 673 (2001).
2.  Liang Fang, Mingrong Shen, and Wenwu Cao, "Effects of post-anneal conditions on the dielectric properties of $CaCu_3Ti_4O_{12}$ thin films prepared on Pt/Ti/SiO$_2$/Si substrates," J. Appl. Phys. 95(11), 6483 (2004).
3.  W. Si, E. M. Cruz, P. D. Johnson, P. W. Barnes, P. Woodward, and A. P. Ramirez , "Epitaxial thin films of the giant-dielectric-constant material $CaCu_3Ti_4O_{12}$ grown by pulsed-laser deposition," Appl. Phys. Lett. 81, 2056 (2002)
4.  M. A. Subramanian, D. Li, N. Duan, B. A. Reisner, and A. W. Sleight, "High dielectric constant in $ACu_3Ti_4O_{12}$ and $ACu_3Ti_3FeO_{12}$ phases," J. Solid State Chem 151, 323 (2000)

COMPARISON OF THE ELECTRICAL BEHAVIOR OF AlN-ON-DIAMOND AND AlN-ON-Si MIS RECTIFYING STRUCTURES

N. Govindaraju[1], D. Das[1], R.N. Singh[1], and P.B. Kosel[2]
[1]Energy and Materials Engineering, University of Cincinnati, Cincinnati, OH 45221-0070
[2]Electrical and Computer Engineering, University of Cincinnati, Cincinnati, OH 45221-0030

ABSTRACT
    Wide bandgap semiconductors such as diamond and AlN are well-suited for high-temperature and high-power electronics applications. This paper reports the results of experiments performed on Metal-Insulator-Semiconductor (MIS) structures consisting of metal—i-AlN—i-diamond—$p^{++}$diamond. It is shown that the combination of i-AlN and i-diamond is critical for achieving diode-like current-voltage (I-V) characteristics. The MIS structures show Schottky behavior with barrier heights of ~0.7 eV and ideality factors of ~17. Analogous device structures consisting of AlN-on-Si show MIS diode-like characteristics for AlN thicknesses below 70 nm with barrier heights of ~0.8 eV. These results indicate that by optimizing process conditions and material properties it is possible to achieve AlN-on-diamond MIS rectifiers for high-temperature in high-power electronics applications.

INTRODUCTION
    Wide-bandgap semiconductors such as diamond and AlN are potentially useful for high temperature electronics since their high bandgaps of 5.5 eV and 6.2 eV, respectively, enable their operation at temperatures well above 300 °C. Furthermore, the dielectric strengths of diamond (10 MV/cm)[1] and AlN (4 – 12 MV/cm)[2] are significantly higher than that of Si (0.3 MV/cm)[1] thereby enabling their operation at high voltage levels.
    This paper reports the development of diamond-based Metal-Insulator-Semiconductor (MIS) diodes for high-temperature electronics. The main device structures discussed here are:
    i.   metal / $p^{++}$ diamond
    ii.  metal / i-diamond (240 nm) / $p^{++}$-diamond
    iii. metal / i-AlN (130 nm) / i-diamond (240 nm) / $p^{++}$-diamond
    Of these three device structures, the third structure (using i-AlN and i-diamond) yielded diode-like current-voltage (I-V) characteristics. Since the growth of diamond thin films is resource and time intensive, a parallel study of this structure was conducted in the AlN-on-Si material system. The results obtained with the AlN-on-Si system indicate that this device structure is viable and provides an opportunity for an AlN-on-diamond based high-temperature rectifier technology.
    To date, a large body of published results has accumulated on polycrystalline and single-crystal diamond-based Schottky diodes[3 - 8]. Since a viable n-doped diamond film has not been established, the majority of the published research has focused on boron-doped p-type diamond. The boron acceptor energy level in diamond resides at 0.37 eV above the valence band edge. This is not a shallow acceptor level, and hence produces an incomplete activation of dopant at room temperature. This in turn implies that the carrier concentration in the diamond is significantly lower than that of the dopant concentration leading to less precise control over the resistivity of the material. Increasing boron concentration eventually leads to degenerate doping and metallic conductivity (at boron concentrations of ~$10^{20}$ cm$^{-3}$). To overcome this drawback, some researchers have tried interposing an undoped diamond layer between the metal electrode

and heavily doped diamond layer[6]. This strategy has met with limited success because of the large intrinsic series resistance present in such device structures. An alternative approach proposed here has been to use a nanometer scale AlN layer as the undoped layer. Previously, Miskys et. al.[9] reported on MBE grown AlN-on-diamond heterojunction photodiodes for UV and blue light emission applications. But, to our knowledge, there is no available literature on AlN-on-diamond MIS device structures for rectifier applications.

DEVICE STRUCTURES

The first structure consisted of a metal in direct contact with heavily $p^{++}$ doped diamond substrate, and the second structure consisted of an undoped diamond layer interposed between the metal contact and the $p^{++}$ doped substrate. The third structure incorporated an additional undoped AlN layer on top of the undoped diamond layer.

The metal-semiconductor contact is the basis for the structures considered in this paper, therefore a brief discussion of its basic physics is appropriate. The details can be obtained in ref. [10]. Figure 1 presents the energy band diagrams for a metal and a p-type semiconductor before and after they are brought into contact. For brevity, only the case for the blocking or rectifying contact is illustrated. When a metal and a semiconductor are brought into contact, the difference in work functions of the metal ($\phi_M$) and the semiconductor ($\phi_S$) causes an initial charge transfer across the interface. When $\phi_M < \phi_S$ (Fig. 1), under equilibrium conditions, this charge transfer results in the establishment of an internal electric field in the semiconductor which effectively prevents further charge flow. The region across which the electric field exists is called the depletion region, since it is depleted of free carriers and contains ionized acceptors. Further, a potential barrier ($\phi_B$) exists which prevents charge flow from the metal to the semiconductor. The potential difference between the surface and the bulk semiconductor material is called the built-in voltage ($V_{bi}$). By applying a forward bias, $V_{bi}$ can be reduced resulting in an exponential increase in current flow. Under reverse bias conditions, $\phi_B$ blocks current flow from the metal to the semiconductor and the depletion width increases to maintain charge balance. The maximum value of the reverse bias voltage that can be supported across the depletion region is determined by the breakdown field of the semiconductor. Therefore, a metal-semiconductor contact can exhibit rectifying I-V characteristics, which in device form, is known as a Schottky diode. The current-voltage expression for a Scottky diode can be shown to be[10]:

$$I = AA^*T^2 \exp\left[-\frac{(q\phi_{B0})}{(kT)}\right] \exp\left(\frac{qV}{nkT}\right)\left\{1 - \exp\left(\frac{-qV}{kT}\right)\right\} \tag{1}$$

where, k is the Boltzmann constant, T the temperature in K, I is the current in amperes, A is the area in cm$^2$, A$^*$ is the Richardson constant, q is the electronic charge, $\Phi_B$ is the barrier height in eV and n is the ideality factor (for an ideal diode n = 1).

Heavy doping of the semiconductor material results in significant changes to the system. The depletion width, under reverse bias conditions, is reduced because the higher volume density of ionized acceptors requires a smaller depletion layer thickness to balance the charge on the metal contact. Under such conditions, the depletion width can become sufficiently small to enable carrier tunneling through the barrier. A significant tunneling current through the barrier can result in a loss of rectifying behavior. The goal of the experiments discussed in this paper was to ameliorate this effect by adding i-diamond and AlN layers between the metal and the $p^{++}$-diamond.

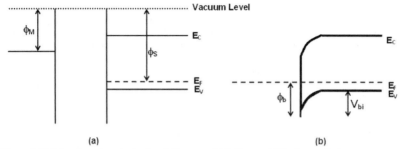

Figure 1.Metal and semiconductor band diagrams (a) before and (b) after contact.

EXPERIMENTAL METHOD

Diamond deposition was carried out in a 1.5 kW (ASTEX, 2.45 GHz) MPCVD system. At the beginning of the fabrication process all samples were thoroughly cleaned by a two-step RCA cleaning procedure to eliminate surface contamination. Standard Si (100) substrates (p-type; 1-20 $\Omega$cm) were used as the host substrate material. All the depositions reported below used a process gas composition of Ar:$H_2$:$CH_4$ in the ratios 60:39:1 respectively at a pressure of 95 Torr. It was found in previous studies[11-13] that it was possible to deposit good quality diamond under these conditions. The temperature of the substrate during deposition was recorded using an infrared pyrometer, which was focused on the substrate surface through a quartz window at the top of the deposition system. Boron doping was obtained by adding 300 ppm of diborane to the gas mixture. The boron doped diamond films considered in this study were 10,000 nm in thickness.

In the case of the AlN-on-diamond devices, thermally evaporated Al ($\sim$ 0.4 $\mu$m thick) served as an etch mask for $SF_6$ plasma etching of the host Si. Following the etching of silicon, the Al masking layer was removed with phosphoric acid. The samples were treated in a 1:1 $H_2O_2$:$H_2SO_4$ solution for half an hour at room temperature to remove residual non-diamond phases from the diamond surface. AlN was deposited by the reactive sputtering of Al with $N_2$ in a RF magnetron sputter system. The AlN sputtering was carried out at a base pressure of 15 mTorr and at a deposition rate of 300 nm/h. All samples utilized Au/Mo bilayer metal electrodes which were circular (of 1 mm diameter). The pads were formed by metal deposition through a shadow mask using a two-step procedure. The first step involved sputter deposition of a 400 nm Mo layer, followed by thermal evaporation of a 100 nm Au layer, to cover the Mo film, through the same shadow mask. The ground plane on each sample was formed by blanket deposition of Au/Mo of similar thickness.

RESULTS AND DISCUSSION

The experimental results for the three types of AlN-on-diamond device structures introduced above presented are here. A separate sub-section highlights the results obtained for the AlN-on-Si devices.

AlN-on-diamond Devices

Table I lists the sample designations and the corresponding device structures for the AlN-on-diamond devices. Figure 2 is a schematic diagram of the device layout for sample BiAlN. This sample consisted of a 10,000 nm (10 μm) uniformly boron-doped diamond layer grown on the surface of a (100) p-type Si host substrate. Subsequently, half of the sample was coated with a 130 nm thick AlN layer as depicted in Fig. 2(a). Current-voltage measurements were carried out on two devices: BiAlN_1 which did not have any AlN deposition on the surface, and BiAlN_2 with 130 nm of AlN deposited on the surface.

Table I. Summary of the sample designations for the AlN-on-diamond devices.

| Sample | Device Structure |
|---|---|
| BiAlN_X (X = 1,2) | Metal / $p^{++}$-diamond |
| BiD_X (X = 1,2,3) | Metal / i-diamond / $p^{++}$-diamond |
| BiDiAlN_X (X = 1...8) | Metal / i-AlN / i-diamond / $p^{++}$-diamond |

Current-voltage (I-V) characteristics of the BiAlN_1 showed linear behavior through the origin indicating that the contacts were ohmic (Fig. 3). As mentioned above, the ohmic contact formation was attributed to the heavily doped diamond layer. Calculation of resistivity from the measured I-V characteristics yielded a value of 261.9 ohm.cm. In the case of the device BiAlN_2, the presence of thin AlN layer on top of the diamond resulted in this device showing some deviation from linearity. These results indicated that it was possible to realize a non-linear current-voltage characteristic using a combination of undoped AlN on doped diamond.

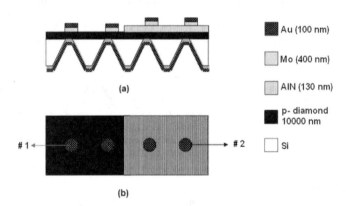

Figure 2. Schematic diagram of device layout for sample BiAlN which did not incorporate an i-diamond layer. Only one half of the sample was coated with AlN (a) front view of the device structure (b) top view of the device structure.

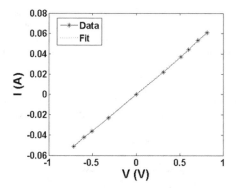

Figure 3. Current voltage characteristics BiAlN_1. Linear behavior is clearly evident.

Since the aim of the experimentation was to demonstrate devices with diode-type I-V characteristics, subsequent device fabrication followed the approach described below. A new sample was fabricated with a 240 nm thickness of undoped diamond grown on top of the 10,000 nm thick boron doped diamond layer (sample designation: BiD). Then a second sample was fabricated with a 164 nm undoped diamond layer grown on top of the 10,000 nm boron doped diamond. A part of the second sample was then coated with 130 nm of AlN through a shadow mask (sample designation: BiDiAlN). For both devices on the second sample the total insulating material thickness was kept below 300 nm.

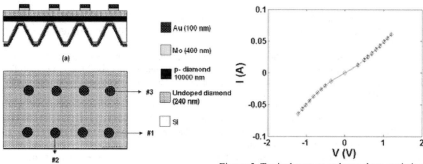

Figure 4. Schematic diagram of sample BiD (no AlN layer present): (a) side view (b) top view.

Figure 5. Typical current-voltage characteristics of device BiD_3 (c.f. Fig. 4 (b)). Non-linear I-V characteristics are evident but it is clear that no rectifying behavior can be observed with a 240 nm thick undoped diamond layer.

Figure 4 is a schematic diagram of the sample BiD. Three devices on this sample were tested (labeled #1 - # 3 in Fig. 4(b)). Figure 5 is a plot of the device characteristics from BiD_3 which is typical of the characteristics observed on this sample. It can be seen from the figure that the deposition of a 240 nm undoped diamond layer produces a non-linear current-voltage characteristics but no rectifying behavior.

From the measurements performed on BiAlN and BiD samples it was concluded that a heavily doped diamond layer coated with a metal electrode (such as Au/Mo) results in an ohmic contact. The incorporation of the thin undoped diamond layer simply results in antisymmetric non-linear I-V characteristic but no diode-like rectifying characteristics. Therefore, the incorporation of a thin AlN layer on top of the undoped diamond layer was the chosen for the third sample structure (BiDiAlN) shown in Fig. 6.

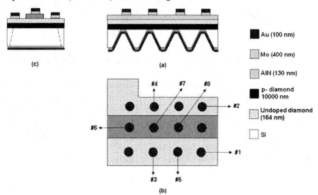

Figure 6. Schematic diagram of the device structures for sample BiDiAlN which incorporates a 130 nm AlN layer on top of the undoped diamond layer: (a) front view (b) top view (c) side view

The top view (Fig. 6(b)) shows that the devices at the center of the sample (# 6 - # 8) incorporated a 130 nm AlN layer in addition to the 164 nm undoped diamond layer. Testing of devices BiDiAlN_1 through BiDiAlN_4 yielded current-voltage characteristics similar to that of sample BiAlN.

Figure 7 shows plots of three devices which incorporate the thin AlN layer on top of the undoped diamond layer (BiDiAlN_6 through BiDiAlN_8). It is clear from the plots that BiDiAlN_7 and BiDiAlN_8 exhibited similar I-V characteristics which now display some rectifying behavior. Since a limited supply of diamond samples was available the emphasis was next placed on extracting useful information from the device characteristics without destroying the devices. Hence, the current and voltage levels were kept low.

The current voltage plots for BiDiAlN_7 and BiDiAlN_8 were fit with the Schottky diode current-voltage relationship (equation (1)) to extract the barrier heights and the ideality factors. The calculated barrier heights for these samples were 0.71 eV and 0.67 eV, and the ideality factors were, 17.3 and 17.1, respectively. Figure 8 shows the results of such a fit BiDiAlN(#7). It can be clearly seen that the Schottky equation fits well with the measured data, other current transport models such as the Fowler-Nordheim tunneling model did not result in a good fit.

The preliminary experiments presented above demonstrated the possibility of achieving rectifying I-V characteristics using i-AlN-on-doped diamond. Since the synthesis and characterization of diamond-based devices was laborious and time consuming, it was decided to test the new rectifying principles on the simpler system of AlN-on-Si.

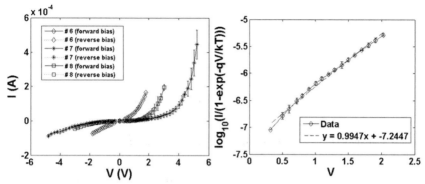

Figure 7. Current voltage characteristics of three devices on sample BiDiAlN. It can be seen that BiDiAlN_7 and BiDiAlN_8 exhibited similar I-V characteristics and hence were subjected to further analysis.

Figure 8. Fit of the current-voltage data to the Schottky model (equation (1) for BiDiAlN_7; $\phi_B$: 0.71 eV and n: 17.3

AlN-on-Si Devices

Electron transport mechanisms in polycrystalline diamond are not well understood because of the complex material issues. For this reason, the i-AlN-on-Si material system was chosen to carry out further investigations of the proposed new device concepts. Silicon was chosen as the substrate material since p-type Si wafers are readily available.

Table II. Sample designations, AlN thickness, the number of devices tested, barrier heights, and ideality factors for AlN-on-Si samples.

| Sample | AlN Thickness | # Devices Tested | $\Phi_B$ (eV) | n |
|--------|---------------|------------------|---------------|-----|
| SiAlN1 | 70 nm | 5 | - | - |
| SiAlN2 | 40 nm | 9 | 0.85 | 63.1 |
| SiAlN3 | 20 nm | 9 | 0.83 | 52.7 |
| SiAlN4 | 20 nm | 3 | 0.77 | 38.8 |
| SiAlN5 | 10 nm | 5 | 0.84 | 20.2 |

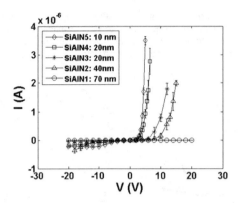

Figure 9. Current-voltage plots for AlN-on-Si devices (c.f. Table II). Rectifying behavior is clearly evident.

Table II lists the sample designations, the AlN layer thicknesses, and the number of devices tested on each sample with Si as the device substrate. The metallization scheme was deliberately chosen to be the same as used for the AlN-on-diamond devices, so that the metal-insulator contact remained the same. Current conduction was observed on samples with AlN thicknesses of 40 nm and below. The measured I-V (Fig. 9) characteristics for these devices were fitted with the Schottky diode model. Table II summarizes the barrier height and ideality factors for all the AlN-on-Si devices. It can be seen from this table that the barrier heights are in the range of 0.8 eV with a variation of only 0.1 eV as compared to the AlN-on-diamond samples. Since AlN is the common dielectric for both these structures, we believe that the AlN film plays a key role in pinning of the barrier height to 0.7-0.8 eV range. Furthermore, the ideality factor improves with decreasing AlN thickness, indicating that the AlN thickness plays an important role in determining the quality of these devices.

These results indicated that it is possible to realize diode-type characteristics with a nanometer scale layer of AlN interposed between the metal contact and the underlying doped semiconductor substrate. In the case of p-type Si substrate, we found that the devices were able to support a large reverse bias voltage (up to 20 V) even with AlN layers as thin as 20 nm as compared to the devices which used heavily doped diamond as the substrate. In the case of doped diamond substrate it may be recalled that the AlN thickness was 130 nm. Even with such a thick coating of AlN, it was found that the current was beginning to increase even under low reverse bias voltages. A possible explanation for this may be associated with the surface roughness of the doped diamond layer. In the 10,000 nm boron doped diamond the average grain size was close to a micron and the surface roughness prevented the formation of a continuous, conformal layer of AlN. This may, therefore, lead to a "patchy" Schottky contact. The ideality factor of ~17 for both devices on sample BiDiAlN indicates that this explanation may be valid. Further studies are proposed to elucidate the exact mechanism for the observed I-V characteristics. The encouraging results from the multiple AlN-on-Si devices lead us to believe that the proposed device concepts are useful.

CONCLUSIONS

Novel AlN-on-diamond device structures consisting of i-AlN, i-diamond, p$^{++}$ diamond were fabricated and tested for exhibiting rectifying current-voltage characteristics. Experiments showed that the metal / i-AlN (130 nm) / i-diamond (240 nm) / p$^{++}$-diamond device structure was the most promising and clearly showed rectifying I-V characteristics. Current-voltage data for these devices were fitted with a Schottky model which yielded barrier heights of ~0.7 eV and ideality factors of ~17.

The proposed device concept was further validated by depositing varying thicknesses of AlN on p-type Si. Results showed that rectifying I-V characteristics were measured for AlN-on-Si devices for AlN thicknesses of 40 nm and below. Data analysis using the Schottky model indicated that the AlN-metal contact may be playing an important role in the observed device behavior. These experiments show that it is possible to develop rectifying device structures using the AlN-diamond materials systems for high temperature electronic applications. Further studies will involve optimizing the process conditions and the layer thicknesses to increase the reverse breakdown voltage, and to reduce the series resistances of these devices.

ACKNOWLEDGEMENTS

The authors are grateful for the support of the National Science Foundation (NSF; Grant No. ECCS0853789) for the present work. Any opinions, findings, conclusions or recommendations expressed in this material are those of the authors and do not necessarily reflect the views of the National Science Foundation.

REFERENCES

[1]Wide Bandgap Semiconductors: Fundamental Properties and Modern Photonic and Electronic Devices, K. Takahashi, A. Yoshikawa, and Adarsh Sandhu, Editors, Springer-Verlag, Berlin, (2007).

[2]T. Adam, J. Kolodzey, C.P Swann, M.W Tsao, and J.F Rabolt, The electrical properties of MIS capacitors with AlN gate dielectrics, Appl. Surf. Sci., 175-176, 428 (2001).

[3]K. Kodama, T. Funak, H. Umezawa, and S. Shikata, The electrical properties of MIS capacitors with ALN gate dielectrics, IEICE Electron. Expr., 7(17), 1246 (2010).

[4]P-N. Volpe, P. Muret, J. Pernot, F. Omnes, T. Tokuyuki, J. Francois, D. Planson, P. Brosselard, N. Dheilly, B. Vergne, and S. Scharnholtz, High breakdown voltage Schottky diodes synthesized on p-type CVD diamond layer, Phys. Status Solidi A, 207(9), 2088 (2010).

[5]A. Vescan, I. Daumiller, P. Gluche, W. Ebert, and E. Kohn, High temperature, high voltage operation of diamond Schottky diode, Diam. Relat. Mater.,7, 581 (1998).

[6]Y. Gurbuz, W. P. Kang, J. L. Davidson, D. V. Kerns, and Q. Zhou, High temperature, high voltage operation of diamond Schottky diode, IEEE T. Power Electr., 20(1), 1 (2005).

[7]D. J. Twitchen, A. J. Whitehead, S. E. Coe, J. Isberg, J. Hammersberg, T. Wikström, and E. Johansson, High-Voltage Single–Crystal Diamond Diodes, IEEE T. Power Electr., 51(5), 826 (2004).

[8]J. E. Butler, M. W. Geis, K. E. Krohn, J. Lawless, S. Deneault, T.M.Lyszczarz, D. Flechtner, and R. Wright, Exceptionally high voltage Schottky diamond diodes and low boron doping, Semicond. Sci. Technol., 18 S67 (2003).

[9]C. R. Miskys, J. A. Garrido, C. E. Nebel, M. Hermann, O. Ambacher, M. Eickhoff, and M. Stutzmann, AlN/diamond heterojunction diodes, Appl. Phys. Lett., 82(2), 290 (2003).

[10]D.K Schroder, *Semiconductor Material and Device Characterization*, (John Wiley & Sons, New York, 1998).

[11]N.Govindaraju, D. Das, R.N. Singh, and P.B. Kosel, High-temperature electrical behavior of nanocrystalline and microcrystalline diamond films, *J. Mater. Res.*, 23(10), 2774 (2008).

[12]R. Ramamurti, V. Shanov, R. N. Singh, S. Mamedov, and P. Boolchand, Raman spectroscopy study of the influence of processing conditions on the structure of polycrystalline diamond films, *J. Vac. Sci. Technol. A*, 24(2), 179 (2006).

[13]R. Ramamurti, Ph.D. Dissertation, University of Cincinnati, Cincinnati, OH (2006).

EFFECT OF NANOCRYSTALLINE DIAMOND DEPOSITION CONDITIONS ON Si
MOSFET DEVICE CHARACTERISTICS

N. Govindaraju[1] , P. B. Kosel[2] , and R.N. Singh[1]
[1]Energy and Materials Engineering, University of Cincinnati, Cincinnati, OH 45221-0070
[2]Electrical and Computer Engineering, University of Cincinnati, Cincinnati, OH 45221-0030

ABSTRACT
    Smooth high thermally conductive diamond films can be used for enhancing the
management of heat in power semiconductor devices. Little appears to be known, however, of
the effects of microwave plasma deposition of smooth diamond films on device performance. To
help fill this void, Si MOSFETs were used as test bed devices for studying such effects in this
study because their technology is well established, and since their performance is sensitive to
process conditions. In this paper effect of deposition of nanocrystalline diamond (NCD), from a
gas mixture of 60% Ar, 39% $CH_4$ and 1% $H_2$, on Si MOSFETs is studied. The transistor current-
voltage characteristics measured after NCD deposition were then compared with those obtained
before deposition. The results generally showed that the MOSFET threshold voltages shifted
after NCD deposition clearly indicating that deposition conditions influence device electrical
characteristics. The implications for the development of an NCD-based thermal management
scheme on semiconductor devices are discussed, and strategies for compensating and/or
neutralizing plasma effects on the devices are considered.

INTRODUCTION
    Power semiconductor electronics technology plays an important role in the continued
evolution of the energy, aerospace, defense, and automotive industries. High current density of
power semiconductor devices requires an efficient thermal management scheme.
    Undoped microcrystalline diamond thin films have been found to have very high
resistivities and to exhibit high thermal conductivities (> 15 W/cmK) when grown to sufficient
thickness, and so have been considered for thermal management applications[1]. However, they
possess very rough surfaces which are not conducive for electronic applications. Therefore, our
research has focused on processing techniques for depositing smooth, highly resistive,
nanocrystalline diamond (NCD) films directly on power semiconductor devices[2]. Our studies
indicate that there are significant thermal performance gains to be achieved by NCD deposition
on top of devices[2].As a first step towards achieving this goal, we have successfully demonstrated
that it is possible to deposit NCD on power semiconductor devices with no visible damage to the
device metallization[2]. The next step in the development of NCD-based thermal management
technology for such devices was to investigate the effects of the microwave excited plasma, used
for the diamond deposition, on device performance. The primary aim of this paper is to describe
preliminary studies on plasma-device interactions using Si metal-oxide-semiconductor field
effect transistors (MOSFETs) as test devices. Si MOSFETs were chosen for this study for the
following reasons.
    i. N-channel Si MOSFET fabrication was available at the University of Cincinnati.
    ii. Si MOSFET technology has evolved over more than four decades and is of central
importance in modern microelectronics technology. Characterization and testing of such devices
is well established, so effects on device performance can be more readily interpreted.

iii. Si MOSFET parameters are very sensitive to process conditions, and therefore are useful for studying the effects of plasmas on device performance.

DEVICE STRUCTURE

Figure 1 is a simple schematic cross section of a typical Si MOSFET. The device consists of a Si substrate, a source, a drain, a gate, and a gate oxide. In simplest terms, the passage of electrons through the MOSFET, from the source to the drain, is modulated by means of the voltage applied to the gate. For a sufficiently large positive gate bias, it possible to modify the surface potential of the semiconductor to such an extent that the minority carriers (electrons) in the p-substrate are attracted to the surface, to form an "inversion layer" (Fig. 2). Under these conditions, when a potential difference is applied between the source and drain, the inversion layer serves as a conduit for current flow between these two terminals. For the device structure illustrated in Fig. 1, the inversion layer consists of electrons and hence this device is called an "n-channel MOSFET" (or an n-MOS transistor). The gate voltage ($V_G$) at which the surface potential in the semiconductor leads to the formation of an inversion layer of sufficiently high conductivity is called the threshold voltage ($V_T$). The MOSFET is considered to be "on" for gate voltages above $V_T$.

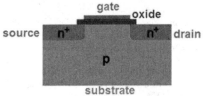

Figure 1. Schematic cross section of a Si MOSFET. The main elements of the MOSFET are shown. The schematic is not to scale.

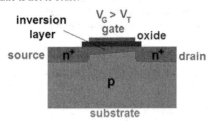

Figure 2. Schematic diagram of a Si MOSFET under operating conditions. The inversion layer thickness is exaggerated for clarity.

For a given gate voltage, $V_G$, exceeding the threshold voltage, $V_T$, increasing the drain voltage ($V_D$) with respect to the source causes in increased current flow (that is, the drain current $I_D$) between the source and the drain. This increasing $I_D$ is nearly linear with increasing $V_D$ until the drain-to-source voltage attains the value $V_{DSat}$ (see Fig. 3). Beyond $V_{DSat}$, $I_D$ remains at a saturated value $I_{DSat}$ until breakdown. For a given $V_D$, increasing $V_G$ results in higher $I_{DSat}$. The overall current-voltage behavior for an n-MOS transistor is illustrated in Fig. 3.

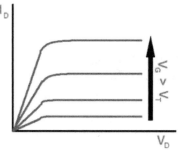

Figure 3. Typical drain current ($I_D$) vs. drain voltage ($V_D$) plots for an n-MOS transistor. The effect of increasing gate voltage is also illustrated.

The formation of the inversion layer in the Si surface under the gate oxide and, therefore, current conduction in the channel are very sensitive to the charge make-up of the interface between the semiconductor and the gate oxide, and the charges present in the gate oxide itself. In general, the threshold voltage dependence on these residual charges can be formulated as[3]:

$$V_T = f(V_{FB}, C_{ox}, N_A)\qquad(1)$$

$$V_{FB} = \phi_{MS} - \frac{Q_{f,ot}}{C_{ox}} - \frac{Q_m}{C_{ox}} - \frac{Q_{it}}{C_{ox}}\qquad(2)$$

where $V_{FB}$ is the flatband voltage in volts, $C_{ox}$ is the oxide capacitance in F/cm$^2$, $N_A$ is the substrate doping concentration in cm$^{-3}$, $\phi_{MS}$ is the work function difference between the gate and the semiconductor in volts, $Q_{f,ot}$ is the combined fixed oxide and oxide trapped charge density in C/cm$^2$, $Q_m$ is the mobile charge in the oxide in C/cm$^2$, and $Q_{it}$ is the interface trapped charge in C/cm$^2$.

The microwave plasma used for diamond film deposition consists of negatively and positively charged species. Consequently, an expected effect of the plasma on a Si MOSFET is the introduction of charged species into the gate oxide (Fig. 4). According to equations (1) and (2), any such introduction of charged species into the gate oxide will result in a shift of the threshold voltage. From the sign of the $\Delta V_T$ shift, the sign of the charge introduced into the gate oxide can be deduced (from equation (2)). This information provides a useful test vehicle for studying the nature of the plasma deposition process and its effects on device operation.

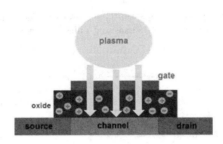

Figure 4. Schematic diagram illustrating a possible mode of plasma-device interaction. The plasma can introduce charged species into the gate oxide thereby altering the current-voltage characteristics of the MOSFET.

EXPERIMENTAL METHOD

The diamond film deposition was carried out in a 1.5 kW (ASTEX, 2.45 GHz) microwave plasma chemical vapor deposition (MPCVD) system. The n-MOS transistors used were fabricated at the University of Cincinnati. The scaled layout of these devices is shown in Fig. 5. The Si MOSFETs did not incorporate any passivation layer making them useful for the study of plasma-device interactions.

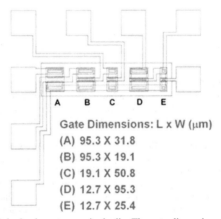

Gate Dimensions: L x W (µm)
(A) 95.3 X 31.8
(B) 95.3 X 19.1
(C) 19.1 X 50.8
(D) 12.7 X 95.3
(E) 12.7 X 25.4

Figure 5. Si MOSFET device layout on a single die. The gate dimensions for different transistors in a single die are also given.

Si device wafers were cleaved into pieces approximately 1 cm x 1 cm in size using a wafer scribing tool. Initial current-voltage characterization of the cleaved Si device samples was carried out before diamond deposition. A novel seeding technique was used to provide uniform and dense diamond seeding layers[2]. Unlike conventional diamond seeding techniques such as ultrasonic agitation and biased enhanced nucleation, the new technique did not result in physical damage to the underlying substrate. NCD was deposited on the device samples at a temperature between 400 °C and 430 °C, at a pressure of 50 Torr, using a gas ratio of 60%:39%:1% (Ar:CH$_4$:H$_2$). Previous studies had shown[4] that this particular gas composition yielded films with superior electrical properties. Three hour depositions (yielding NCD thicknesses < 1 μm) were chosen for the following reasons. Thick NCD films were avoided because diamond is an inert material, and etching to make contact to the metal probe pads would complicate the interpretation of data. Access to contact pads when using thin NCD films (< 1 μm) was easily realized by using sharp-pointed probes that easily penetrated through the NCD layer. In this way, current-voltage characterization was performed on the devices after NCD deposition and yielded the results presented below.

RESULTS AND DISCUSSION

Figure 6 shows a micro-Raman spectrum (at 50X; 532 nm; 25 mW nominal power; Nicolet Almega, Thermo Fisher Scientific, Inc, MA, USA) taken from a Si device sample after the NCD film deposition. The first order Raman peak at 1332 cm$^{-1}$, characteristic of diamond, is clearly visible in Fig. 6.

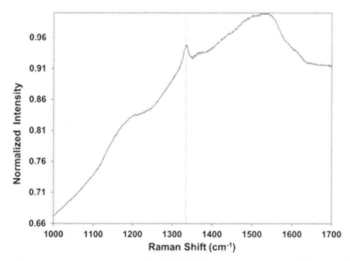

Figure 6. Micro-Raman spectrum on a Si-MOSFET device sample after NCD deposition. The first order Raman peak at 1332 cm$^{-1}$ is evident, indicating that diamond deposition is taking place.

Subsequently the I$_D$-V$_D$ characteristics for the Si MOSFET were measured as shown in Fig. 7, and overlaid on those obtained before NCD film deposition. The continuous lines represent data measured before NCD deposition, while the dashed lines represent data measured after deposition. A significant change in the current-voltage characteristics is evident.

Specifically, the data indicates that the threshold voltage has decreased, and that for a given drain voltage ($V_D$) a higher drain current ($I_D$) flows through the device after diamond deposition.

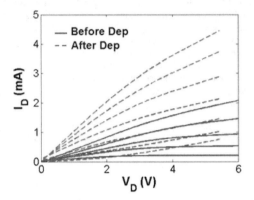

Figure 7. Typical measured current-voltage characteristics for a Si-MOSFET before and after NCD deposition. The threshold voltage lowering and increased drain current flow are obvious.

Table I summarizes the results of threshold voltage changes obtained for multiple devices on different die. It is clearly seen that there is a consistent shift to lower threshold voltages for all devices after diamond deposition. Furthermore, the average threshold voltage reduction of ~1.5 V represents a significant change in device characteristics. The lowering of the threshold voltage indicates that the charges incorporated into the gate oxide are positive (equation (2)).

Table I. Summary of threshold voltage data collected before and after NCD deposition. The data corresponds to the device structure "C" (c.f. Fig. 5) on different die.

| Die | $V_T$ (V) | | $\Delta V_T$ (V) |
|---|---|---|---|
| | Before Dep. | After Dep. | |
| A16 | 3.96 | 3.01 | 0.95 |
| C16 | 4.82 | 2.13 | 2.69 |
| E16 | 4.21 | 3.11 | 1.1 |
| F17 | 4.10 | 2.65 | 1.45 |
| G16 | 3.97 | 2.53 | 1.44 |
| Avg. | $4.21 \pm 0.36$ | $2.69 \pm 0.39$ | $-1.53 \pm 0.69$ |

The above study has proved to be a useful vehicle for investigating the basic processes involved in plasma-device interactions during NCD deposition. Further studies are focused on evaluating the nature of the charge incorporated into the gate oxide, and also possible strategies for nulling out the effects of the plasma on the device performance. Two ways currently under study are to: (1) deposit a thin barrier layer on top of the device by non-plasma methods, and (2) use a high temperature anneal.

CONCLUSIONS

Si MOSFETs were used as test bed devices to study plasma-device interactions during NCD deposition. Results indicate that the plasma has a direct and significant influence on the device performance. In particular, NCD deposition results in the incorporation of positive charge into the gate oxide, thereby lowering the threshold voltage of these devices. Further research is focusing on developing a deeper understanding of the nature of this interaction, and will aim at developing strategies for annulling such effects. This work will enhance the understanding for new and efficient thermal management technology for wide bandgap semiconductor electronics with NCD layers.

ACKNOWLEDGEMENTS

The authors are grateful for the support of the National Science Foundation (NSF; Grant No. ECCS0853789 and CBET-1133516) for the present work. Any opinions, findings, conclusions or recommendations expressed in this material are those of the authors and do not necessarily reflect the views of the National Science Foundation.

REFERENCES

[1]J.E.Graebner, S. Jin, G.W. Kammlott, J.A. Herb, and C.F. Gardinier, Large anisotropic thermal conductivity in synthetic diamond films, *Nature*, **359**, 401-03 (1992).

[2]N. Govindaraju and R.N. Singh, Processing of nanocrystalline diamond thin films for thermal management of wide-bandgap semiconductor power electronics, *Mater. Sci. Eng. B*, **176(14)**, 1058-1072 (2011).

[3]D.K Schroder, Semiconductor Material and Device Characterization, (John Wiley & Sons, New York, 1998).

[4]N.Govindaraju, D. Das, R.N. Singh, and P.B. Kosel, High-temperature electrical behavior of nanocrystalline and microcrystalline diamond films, *J. Mater. Res.*, **23(10)**, 2774-86 (2008).

STUDY OF THE DIFFUSION FROM MELTED ERBIUM SALT AS THE SURFACE-MODIFYING TECHNIQUE FOR LOCALIZED ERBIUM DOPING INTO VARIOUS CUTS OF LITHIUM NIOBATE

Jakub Cajzl, Pavla Nekvindova, Blanka Svecova, and Jarmila Spirkova
Department of Inorganic Chemistry, Faculty of Chemical Technology, Institute of Chemical Technology, Technicka 5, 166 28 Prague, Czech Republic

Anna Mackova, Petr Malinsky, and Jiri Vacik
Nuclear Physics Institute, Academy of Sciences of the Czech Republic, v.v.i., 250 68 Rez, Czech Republic
Department of Physics, Faculty of Science, J. E. Purkinje University, Ceske mladeze 8, 400 96 Usti nad Labem, Czech Republic

Jiri Oswald
Institute of Physics, Academy of Sciences of the Czech Republic, Cukrovarnicka 10, 162 53 Prague, Czech Republic

Andreas Kolitsch
Helmholtz-Zentrum Dresden Rossendorf, 01314 Dresden, Germany

ABSTRACT
The results of the localized doping of erbium into lithium niobate (LN), mainly via diffusion from melted erbium salt (TD – thermal diffusion), are presented. Two different temperatures, 350 and 600 °C, as well as commonly used and specially cut LN wafers were applied. The samples were characterized by Rutherford Backscattering Spectroscopy (RBS) for the erbium concentration depth profiles and by Photoluminescence Spectroscopy for the emission around 1530 nm. The samples revealed thin erbium-doped layers that contained from 7 to 24 at. % of erbium. The luminescence without the post-diffusion annealing appeared in all of the LN cuts. As expected, when a higher temperature of TD was used, the luminescence intensity increased. However, the big differences in the luminescence intensity were found between the various LN cuts. A higher value of luminescence intensity was always detected in the $Y_\perp$ cut <10–14>, which is perpendicular to the crystal cleavage plain. To clarify the different penetration of the crystallographic orientation of LN for incoming erbium ions as well as the mechanism of the thermal diffusion process, the erbium site in the LN structure was studied using a combination of RBS/channeling and Neutron Depth Profiling (NDP) methods.

1.    INTRODUCTION
Erbium-doped lithium niobate ($Er:LiNbO_3$, Er:LN) is, thanks to its suitable properties and mainly because of the possibility of transmission of the signal in the third telecommunication window, a very attractive material for photonics devices[1-3]. However, the doping of LN with Rare-Earth elements is very demanding, so that most of the devices are formed from bulk-doped LN in which the waveguiding layer is consequently created by high-temperature diffusion of titanium or zinc[4,5]. Therefore, the creation of an optically active waveguiding thin layer by an erbium localized doping into the LN surface is a tempting question to research. Actually, such an approach has already been studied and the localized doping of erbium from metal erbium or an erbium oxide film is currently being used[6]. This approach has also become our long-term task and within it we are going to compare

techniques and eventually suggest a feasible approach for the localized doping of erbium into LN leading to the construction of a real thin-film optical amplifier. In one of our previous papers[7,8], we reported on novel technology of the localized LN doping using a moderate-temperature diffusion of $Er^{3+}$ from the melt of nitrates. We have tested this approach on several different crystallographic cuts of LN and found that all of the doped cuts revealed the emission at 1.5 μm. The amount of the in-diffused erbium depended strongly on the crystallographic orientation of the pertinent cuts. Similar results were also obtained for diffusion from the melts of sulfates performed at the higher temperature; this experiment was originally done by Sada et al.[9]. As expected, the crucial factor for a successful doping is the temperature of the diffusion, and it is true not only for the rate of the diffusion but also for the mechanism involved. Several interesting questions still remain unanswered, such as why the intensity of the luminescence is different in different crystallographic cuts after the thermal diffusion and what the basic processes occurring during the diffusion that affect the sites of the incorporated erbium are. The basic problem is whether the process is based on the exchange of erbium ions for lithium ions.

In this study, we are going to report on our experiments on the doping of erbium from the melts containing erbium salts conducted at two different temperatures into commonly used and specially cut LN wafers. The samples were characterized by luminescence measurement and by ion-beam techniques such as RBS (Rutherford Backscattering Spectroscopy) for the amount of the incorporated erbium. Using the RBS/channeling method and NDP (Neutron Depth Profiling), we were able to study the mechanism of the thermal diffusion process.

2.    EXPERIMENT

For the experiments, we used wafers of LN with various crystallographic orientations (pulled by the Czochralski method) supplied by Crytur Turnov (Czech Republic). The wafers were oriented and labeled as follows: standard X cuts <11–20> and Z cuts <0001> and specially executed Y cuts with respect to the cleavage plane of the LN crystal, namely a parallel $Y_{II}$ cut <10-12> and perpendicular $Y_{\perp}$ cut <10-14>. All of the wafers were ground on both sides and polished on one side to optical quality. The dimensions of the wafers were 12 × 7 mm with a thickness of 0.7 mm. Before the experiments, the wafers were treated by ultrasound-assisted cleaning in isopropanol.

For the doping of erbium, two erbium salt-containing multi-component melts were used: a two-component nitrate melt and a four-component sulfate melt (see Table I). The compositions of the melts differed according to thought temperature of the diffusion. The actual composition was based on our previous experiments[7,8] on finding suitable melts that would allow for doing the diffusion experiments mechanisms of that could have differed each from other. The criteria for the suitable melts were, besides a feasible and user friendly experimental procedure, mainly that the melts would not make any harm to the single-crystalline surfaces of the samples. Unfortunately it was found that it was not possible to find any melts that would fulfill the expected criteria for both temperatures of the diffusion; it means that the actual compositions of the melts differed not only in the particular components but also in the actual content of erbium. The amount of erbium nitrate was as high as possible but simultaneously below the level that would allow for the separation of $Er_2O_3$ that may occur at higher temperatures. Any diffusion of the large potassium cation did not expect. The doping from the nitrate melts carried out at 350 °C for 168 hours. The sulfate melt consisted, according to[9], of $Li_2SO_4$, $Na_2SO_4$ and $K_2SO_4$ (100 wt. %), into which 1 wt. % of $Er_2(SO_4)_3$ was added. The doping from this melt was carried out at 600 °C for 40 hours. Therefore, two basic sets of the samples were prepared and in order to ensure a good reproducibility of the results the experiments using both of the melts were repeated.

The doping was done in an oven using platinum crucibles. To ensure the homogeneity of the doping media, the melts were heated for at least 3 hours prior to the doping process.

For the recovery of the LN structure, the as-doped samples were annealed for 5 hours at 600 °C in oxygen.

Table I. The erbium-melts composition

| Melt | Component * | Content [wt. %] |
|---|---|---|
| Nitrates | $KNO_3$ | 90 |
| | $Er(NO_3)_3$ | 10 |
| Sulfates | $Li_2SO_4$ | 70.7 |
| | $Na_2SO_4$ | 9.9 |
| | $K_2SO_4$ | 19.4 |
| | $Er_2(SO_4)_3$ | 1.0 |

*As some of the components, e.g. erbium nitrate, lithium or erbium sulfates, occur as hydrated salts, the compositions of the melts were evaluated for the anhydrous ones.

The concentration profiles of the incorporated erbium ions were studied by RBS. The analysis was performed at a Tandetron 4130 MC accelerator using a 2.0 MeV He$^+$ ion beam. The backscattered He$^+$ ions were detected at a laboratory angle of 170°. The collected data were evaluated and transformed into concentration-depth profiles using the GISA 3 computer code[10]. The RBS/channeling analysis was performed to study the structural changes in the surface layer containing Er and to obtain information on the Er position in crystal after thermal diffusion. The RBS/channeling measurements were performed using a 1.7 MeV He$^+$ beam from the Van de Graaff accelerator in the Helmholtz Zentrum.

The photoluminescence spectra of the Er doped samples were collected within the range of 1440–1600 nm at room temperature. A pulse semiconductor laser POL 4300 emitting at 980 nm was used for the excitation of the luminescence. The luminescence radiation was detected by a two-step-cooled Ge detector J16 (Teledyne Judson Technologies). To scoop the specific wavelengths, a double monochromator SDL-1 (LOMO) was used. For the evaluation, all of the luminescence spectra were transformed to the base level and after the abstraction of the baseline the normalizing was done with help of reference samples.

The lithium concentration depth profiles in the prepared samples were measured by Neutron Depth Profiling (NDP). The method is based on a reaction of a thermal neutron with $^6$Li: $^6$Li (n,α) $^3$H. The fabricated samples were irradiated with a thermal neutron beam from a 6 m long neutron guide (the neutron intensity was $10^7$ neutrons.cm$^{-2}$.s$^{-2}$) and the charged reaction products were recorded by means of a Si(Au) surface barrier detector. The accuracy of the NDP method is 5% of the concentration value of Li, $c_{Li}$, the depth resolution is 10 nm. The natural abundance of the "NDP active" $^6$Li isotope is 7.5%, but in the actual samples the $^6$Li/$^7$Li ratio may significantly vary (e.g., due to the artificial depletion of the $^6$Li isotope from the original natural materials). Thus, to avoid the uncertainty induced by this variation, for some considerations we rely on the relative changes of $c_{Li}$ rather than their absolute values.

3.    RESULTS

According to our previously conducted research[7,8], we assumed that the amount of the incorporated erbium in the doped thin-surface layers will be strongly affected mainly by the temperature of the doping but also by the type of the different crystallographic cuts. Four types of various LN cuts were doped at two different temperatures, i.e., 350 °C and 600 °C. The actual temperatures were selected on the basis of previous experiments showing that at 350 °C the migration of lithium ions can be expected and at 600 °C the damaged LN crystal structure might be recrystalized[11] (the experimental conditions and composition of the melts used for the doping have been mentioned above in Chap. 2). The crucial property that was followed was the intensity of the luminescence at 1.5 μm, which was measured by luminescence spectroscopy in the reflection arrangement. The RBS method was used to determine the integral amount and depth distribution of the incorporated erbium, while the NDP method was a useful tool to follow the migration of lithium during the doping process. With the help of RBS/channeling, we were able to evaluate the structural changes after the TD and the amount of Er in the interstitial position of the LN structure.

## 3.1 LUMINESCENCE

In this chapter, we are going to compare the intensities of the luminescence bands occurring between 1400 nm and 1600 nm, with special attention to the band at 1530 nm, as it is particularly important for the potential utilization of the pertinent material in photonic devices. The samples doped at 350 °C from the nitrate melt revealed the maximal 1530 nm band intensity around 200 a.u. being the highest in the $Y_\perp$ cut perpendicular to the cleavage plane and decreasing in the order: $Y_\perp \rightarrow Z \rightarrow Y_{II} \approx X^{7,8}$.

The samples doped from the sulfate melt revealed a much higher intensity of the luminescence at 1530 nm; the intensity of the best ones reached up to 4000 a.u. (i.e., 20 times higher than the highest intensity of luminescence observed in samples fabricated using nitrate melt). Compared with TD from the nitrate melt, the $Y_\perp$ cut also revealed a high intensity of 1530 nm luminescence. Surprisingly, intensive luminescence was detected at the Z cuts, which was even higher than that at the $Y_\perp$ cut (see Fig 1b).

In the spectra of the samples doped from the sulfate melt, a distinctive band at 1490 nm was also found, which can be attributed to the irradiative transition $^4I_{15/2} \rightarrow {}^4I_{13/2}$ (like the emission at 1530 nm) and is a consequence of a fine splitting of the $Er^{3+}$ levels $^4I_{15/2}$ and $^4I_{13/2}$ in the crystal field of the LN. This band is much deeper in the samples prepared by doping from the sulfate melt.

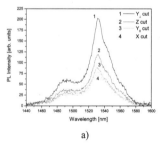 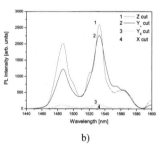

a)          b)

Fig. 1 The luminescence spectra of the erbium-doped LN samples prepared from the melt of a) nitrate melt (168 hrs at 350 °C), b) sulfate melt (40 hrs at 600 °C)

## 3.2 ERBIUM-CONCENTRATION PROFILES AND STRUCTURAL CHANGES IN LN

The concentration-depth profiles of the erbium incorporated in all of the LN samples prepared by doping from the nitrate melt are shown in Fig. 2a. Obviously, the surface concentration of erbium differs in the particular samples, but it does not exceed the range from 2 at. % to 7 at. %. The shape of the concentration-depth profiles showed an exponential decrease of the erbium content towards depths of approximately 40 nm under the surface of the samples. The highest concentration of erbium was found in the Z and $Y_\perp$ cuts, which also had very similar shapes of the depth-concentration profiles. The lowest concentration of erbium was found in the $Y_{II}$ cut.

The erbium doping from the sulfate melt was performed in all four types of cuts. The obtained concentration-depth profiles of erbium differed when compared with samples conducted in nitrates and are displayed in Fig. 2b. In Fig. 2b it is evident that the concentration profile of erbium is maximal at the surface and is decreasing towards the LN depth like the profiles observed with samples fabricated in nitrate melt. However, it is possible to see clearly the very high concentration of erbium reaching 24 at. %, which means that almost one quarter of the surface (in terms of atomic concentration) is occupied with erbium. Such a high concentration of erbium even changed the appearance of the samples that showed a coloring similar to that of the interference images. The experiments were rather demanding, and therefore, to be sure that the results would be fully reproducible, we repeated the experiments with sulfate melt. In all of the prepared samples, the surface concentration of erbium was rather different in particular LN cuts (ranging from 12 at.% to 24 at.%). The highest concentrations of erbium were in all of the sets of the samples always found in the $Y_\perp$ and the Z cuts. The depths of the doped layers reached 70 up to 100 nm.

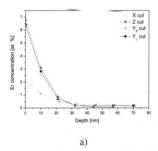
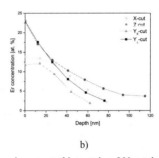

a)                                                              b)

Fig. 2 The depth-concentration profiles of the erbium incorporated into various LN cuts by diffusion from the a) nitrate melt (168 hrs at 350 °C), b) sulfate melt (40 hrs at 600 °C)

An RBS/channeling analysis would enable us to determine the amount of Er atoms incorporated in the interstitial sites looking along the main axis of each crystallographic cut. However, the prepared erbium-doped samples, both from the nitrate and sulfate melts, were so disordered as to make directing the channeling and evaluating the positions of the erbium atoms in this way impossible. The number of disordered atoms is comparable to an amorphous structure; it means that the thermal diffusion causes a high modification of the crystalline structure. In the RBS random spectra, we can observe Nb depletion, see Figure 3a. The depletion of Nb in the surface layer was observed in all of the cuts of LN with Er incorporated by diffusion from the sulfate melt. The depleted amount of Nb was about 5–7 at. %.

Results that are even more interesting were obtained after annealing the sample sets doped at 600 °C from the sulfate melt (the experimental conditions of annealing were mentioned above in Chapter 2). In these samples, the erbium-surface concentrations decreased and erbium atoms were redistributed to a depth. Moreover, differences between the particular

LN cuts appeared. The best movement of erbium atoms was observed in the $Y_\perp$ cut, where the erbium-containing layer reached up to 0.5 μm and the erbium surface concentration decreased to 18 at.% (Fig. 3). The other LN cuts currently revealed the 10 at.% concentration of erbium even in 0.1 μm depths.

Such samples also allowed the directing of the channeling spectra so that the position of the erbium could be evaluated. The samples after the annealing procedure exhibited a higher level of ordered structure as compared to the non-annealed samples, which enabled us to find the channeling direction and compare the amount of Er incorporated in the substitutional and interstitial positions. It is obvious that 52 at. % of Er atoms were located in the interstitial positions in the $Y_\perp$ cut and the remaining amount of Er atoms was placed in the substitutional position of this cut. This phenomenon was observed only in the mentioned cut, not in the other cuts investigated.

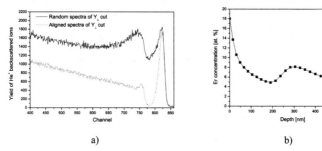

a)                                                           b)

Fig. 3 The RBS channeling spectra of $Y_\perp$ cut of LN fabricated by TD in sulfate melt (a), calculated erbium concentration-depth profile (b)

3.3 LITHIUM CONCENTRATION PROFILES

All of the annealed samples were measured using the NDP method to determine the changes of the concentrations of lithium in the surfaces of the samples. This method is a suitable tool as we know from our previous experience, because the measurement does not make any changes in the distribution of the lithium atoms. The results of the measurement are shown for the samples doped at lower temperature (from the nitrates melt) in Fig. 4a and those doped at higher temperature (from the sulfates melt) in Fig. 4b.

The depth-concentration profiles displayed in Fig. 4a for the samples doped at 350 °C clearly show the depletion of lithium from the surface, thus increasing its concentration towards the subsurface areas of the samples. The percentual depletion of lithium was evaluated from the integral-area ratio above the depth profiles of lithium reaching down to 0.5 μm (see Table II). From the comparison of those areas, it can be seen that on the surfaces of the samples doped at 350 °C about 20 at. % of lithium is missing. Table II also shows that the depletion of lithium is bigger in the X and $Y_\perp$ cuts while the lowest depletion is in the Z cut.

The samples doped at higher (600 °C) temperatures are dramatically different. The concentration-depth profiles show that the depletion of lithium is located in a very narrow area below the surface of the samples. At a depth of 10 nm under the surface, the concentration of lithium in all of the samples is practically the same as in the virgin LN samples. Substantial differences were noted between particular cuts: according to the integral amount the biggest depletion occurred in the X cut while in the other cuts the depletion was much smaller. The depth to which the depletion of lithium (10 nm) reached does not in any case correspond with the depth where erbium was still detected (up to 100 nm).

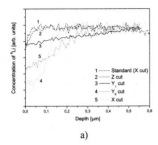
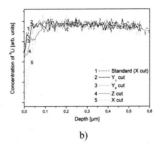

a)

b)

Fig. 4 The depth-concentration profiles of lithium in various LN cuts doped with erbium from the melts of a) nitrates (168 hrs at 350 °C), b) sulfates (40 hrs at 600 °C).

Table II. The percentual depletion of lithium calculated from the integral amount of the lithium concentration profiles

| | Integral amount [atoms/cm²] | Percentual depletion of lithium [%] |
|---|---|---|
| Nitrates : LN $Y_{II}$ cut | $2.28 \times 10^{16}$ | 26.4 |
| Nitrates : LN $Y_{\perp}$ cut | $1.41 \times 10^{16}$ | 16.3 |
| Nitrates : LN X cut | $2.33 \times 10^{16}$ | 27.0 |
| Nitrates : LN Z cut | $2.93 \times 10^{15}$ | 3.4 |
| | | |
| Sulfates: LN $Y_{II}$ cut | $3.03 \times 10^{15}$ | 3.5 |
| Sulfates : LN $Y_{\perp}$ cut | $2.44 \times 10^{15}$ | 2.8 |
| Sulfates: LN X cut | $9.27 \times 10^{15}$ | 10.7 |
| Sulfates: LN Z cut | $5.52 \times 10^{15}$ | 6.4 |

## 4.  DISCUSSION

The presented results provide evidence that it is in principle possible – using purely thermal diffusion of erbium from erbium-containing salts – to prepare a very thin optical layer on the surface of the LN crystal that reveals luminescence at 1530 nm. The thickness of layers prepared in this way was maximally 0.1 μm, which is, in terms of the experimental conditions used, in good agreement with the results of Sada[12] and Caccavale[13]. As expected, the crucial factor that influenced the process of thermal diffusion was temperature, which affected not only the rate of the penetration of the diffusing erbium ion but also the mechanism of how it occurred. Generally, on the bases of the results obtained by SIMS, it can be argued that the thermal diffusion of erbium is in fact an ion exchange of the erbium ions from the melt for the lithium ions from the surface of the LN crystals[1]. This mechanism is in the case of diffusion from the melt of nitrates likely to occur at lower temperatures, where the thickness of the in-diffused layer corresponds approximately with the thickness of the lithium-depleted layer.

It is also important to bear in mind that the process of diffusion from the nitrates-melt is much slower and takes a very long time, so that, in principle, ion exchange may also take place. This has also been confirmed in our earlier papers[7,8], where we gave evidence that erbium in-diffusion was always accompanied by the migration of lithium through the structure of the LN. The question is whether the erbium ions would occupy the sites freed by lithium ions. In the case of samples with the highly disordered surface layer there is no sense to talk about the Er position, because this position is random in the randomly ordered structure.

A slightly different situation arises at the diffusion performed at higher (600 °C) temperatures. According to our results, it is unlikely that an exchange of erbium for lithium ions occurs there. One should remember that the time of the diffusion is much shorter and lithium is already present in the melt used for the diffusion of erbium. Even though the amount of erbium incorporated into the LN surface is much higher (as compared with the lower temperature approach), depletion of lithium is practically zero; this indicates that the mechanism of the higher-temperature diffusion will be different. According to[11], the structure of the LN crystal at such a temperature may partly recrystallize. This idea leads us to believe that after the incorporation of erbium ions into the LN structure a balancing of the charges may occur in such a way that erbium ions (during the recrystallization) may occupy the positions of niobium while the amount of lithium remains unchanged. This hypothesis is verified by the highest intensities of the 1530 nm luminescence bands, which are evidence of a better ordering of the erbium surrounding and moreover the depletion of lithium atoms is very low. During the post-diffusion annealing the recrystallization process follows on – erbium ions from the interstitial sites move to the substitutional sites. The differences between the various cuts were found: in the $Y_\perp$ cut more than 52% of incorporated Er after the annealing procedure were in the interstitial sites while in the Z cut the amount of Er in interstitial sites remained 93 - 99% as to the random RBS/channeling spectrum.

Concerning the mechanism of the thermal diffusion from the point of view of different single-crystalline lithium niobate cuts, the importance of the crystallographic orientation of the cuts with respect to the cleavage plane is patently evident. The greatest amount of incorporated erbium as well as the highest intensity of the 1530 nm were always connected with the specially designed $Y_\perp$ cut <10–14>. If one accepts the concept of an ion exchange process of the erbium incorporation, then for comparison one can use the well-known method of proton exchange (PE or APE) for the fabrication of optical waveguides in lithium niobate in the standard X, Y and Z cuts of LN. The rate of proton exchange is always higher in the X cuts than in the Z cuts; this has also been proven in our research concerning the migration of lithium through the LN structure and the greatest depletion of lithium in the X cuts [see Fig.4]. Then, why was the highest erbium concentration not found also in the X cut? Actually, the opposite is true. In light of this, the ion-exchange mechanism of erbium incorporation does not seem likely.

5. CONCLUSION

The thermal diffusion of erbium at two temperatures into commonly, as well as specially cut LN wafers, was performed. The thin layers which were prepared contained different amounts of erbium, i.e., 7–24 at. % depending on the crystallographic orientations of the wafer surfaces at a depth of less than 100 nm. As expected, the intensity of the luminescence corresponds well with the amount of incorporated erbium. The highest luminescence intensity therefore was even more obvious when the higher temperature was applied. Moreover, the positive effect of annealing on the layer containing erbium was also verified. Using the RBS/channeling and NDP methods, the mechanism of TD was studied and the following facts became evident:

i) For the mechanism of the process of incorporating erbium into the LN structure it was found that rather than the amount of erbium in the melt used for the diffusion it is the temperature of the process that is strongly determinative. While the mechanism of the ion exchange is more likely at 350 °C (lithium-atom depletion was detected in the surface layer), at a 600 °C temperature the mentioned mechanism was not proved. It is interesting that despite the fact no depletion of lithium occurred (probably because the high content of it in the used sulfates melt, it evidently occurred the in-diffusion of erbium. It indicates that the mechanism of incorporation of erbium via ion exchange is under those actual conditions unlikely.

ii) The ability of erbium to penetrate into the LN structure is clearly influenced by the sample surface crystallographic orientation. The highest erbium content and the highest luminescence intensity at 1530 nm were always observed in the $Y_\perp$ cut <10–14> perpendicular to the cleavage plane.

iii) After TD, the position of the erbium atoms was not possible to determine using RBS/channeling spectroscopy since the majority of the atoms have been deflected from their positions and therefore most of the erbium atoms are located in the interstitial positions. The subsequent annealing caused a recrystallization of the structure, erbium penetration deeper into the LN substrate and its occupation of the substitution positions.

REFERENCES

1K.K. Wong, Properties of Lithium Niobate, INSPEC (2002) London.
2L. Arizmendi, Phys. Stat. Sol. (a) 201 (2004) 253.3W.
3Sohler et al., IEICE TRANS. ELECTRON E88–C (5) (2005) 990.
4 G. Singh, S. Gupta, S. Bothra, V. Janyani and R. P. Yadav, Proc. SPIE 8069, (2011) 80690W.
5E. Cantelar, R. Nevadö, G. Maitín, J.A. Sanz-García, G. Lifante, F. Cusso, M.J. Hernandéz, P.L. Pernas, J. Lumin. 87-89 (2000) 1096.
6I. Bauman, R. Brinkmann, M. Dinand, W. Sohler, L. Beckers,Ch. Buchal, M.Fleuster, H. Holzbrecher, H. Paulus, K.-H. Muller, Th.Gog, G. Materlík, O.Eitte, H.Stolz, W.von der Osten, Appl. Phys. A 64, (1997) 33.
7J. Hradilová, P. Kolářová, J. Schrofel, J. Čtyroký, J. Vacík, J. Peřina, Proc. SPIE 2775 (1996) 187.
8V.Peřina, J.Vacík, V.Hnatovitz, J. Červená, P. Kolářová, J. Špirková-Hradilová, J. Schröfel, Nucl. Instrum. Methods Phys. Res., Sect. B 139 (1-4), (1998), 208–212.
9C. Sada, F. Caccavale, F. Segato, B. Allieri, L.E. Depero, Optical Materials 19 (1) (2002) 23–31.
10J. Saarilahti, Nucl. Instr. Methods Phys. Res. B 64 (1994) 734.
11A. Meldrum, J. Nucl. Mater. 300 (2002) 242.
12C.Sada et al., Appl. Phys. Lett. 72 (1998) 3431.
13F.Caccavale, C. Sada, F. Segato, N.Yu. Korkishko, V.A. Fedorov, T.V. Morozova, J. Non. Cryst. Solids 245 (1999) 135.

ACKNOWLEDGEMENT

The authors acknowledge support from the following grants: Czech Science Foundation 106/10/1477 and 106/09/0125, TA01010237. This work has also been supported by the European Community as an Integrating Activity 'Support of Public and Industrial Research Using Ion Beam Technology (SPIRIT)' under EC contract no. 020 and 227012.

ACOUSTIC WAVE VELOCITIES MEASUREMENT ON PIEZOELECRTIC CERAMICS TO EVALUATE YOUNG'S MODULUS AND POISSON'S RATIO FOR REALIZATION OF HIGH PIEZOELECTRICITY

Toshio Ogawa and Takayuki Nishina
Department of Electrical and Electronic Engineering, Shizuoka Institute of Science and Technology, 2200-2 Toyosawa, Fukuroi, Shizuoka 437-8555, Japan

ABSTRACT
    The dependences of longitudinal and transverse wave velocities, Young's modulus and Poisson's ratio on the composition and firing processes were investigated in hard and soft PZT, alkali niobate and alkali bismuth titanate ceramics. The fluctuation of the velocities in transverse wave was smaller than the ones in longitudinal wave by an ultrasonic thickness gauge with high frequency pulse oscillation. The effect of reducing pores in ceramics fired in oxygen atmosphere on the elastic constants works like the increase of the longitudinal wave velocity and Poisson's ratio, and the decrease of the transverse wave velocity and Young's modulus. It was found the dependences of the longitudinal and transverse wave velocities, Young's modulus and Poisson's ratio on the firing position of PZT ceramics in sagger. Furthermore, it was clarified that there are compositions with high piezoelectricity in the cases of (1) low Young's modulus and high Poisson's ratio (PZT type) and (2) low Young's modulus and low Poisson's ratio (PbTiO$_3$ type) in lead-free ceramics.

INTRODUCTION
    Elastic constants of piezoelectric ceramics were basically evaluated by impedance responses on frequency of piezoelectric resonators with various relationships between DC poling directions and vibration modes[1-4]. Therefore, it needs all kinds of resonators with different shapes[1], and furthermore, with uniform poling-degree in spite of different poling-thicknesses such as 1.0 mm for plate resonators (for example, dimensions of 12 mm length, 3.0 mm width and 1.0 mm poling-thickness) and 15 mm for rod resonators (dimensions of 6.0 mm diameter and 15 mm poling-thickness). While applying a DC poling field of 3.0 kV/mm, 3.0 kV DC voltage must be applied to the former sample and 45 kV for the later sample. There was no guarantee to obtain uniform poling-degree between two samples even through the same poling field of 3.0 kV/mm. In addition, it is difficult to measure in the cases of as-fired ceramics, namely before poling ceramics, or weak polarized ceramics because of none or weak impedance responses on frequency.
    On the other hand, there was a well-known method to measure the elastic constants by pulse echo measurement[5,6]. Acoustic wave velocities in ceramics were directly measured by this method. However, since the frequency of pulse oscillation for measurement was generally below 5 MHZ, it needs to prepare the rod samples with thickness of 10-20 mm in order to guarantee accuracy of the velocities. Moreover, it is unsuitable for measuring samples such as disks with ordinary dimensions (10-15 mm diameter and 1.0-1.5 mm thickness) and with many different compositions for piezoelectric materials R & D.
    We developed a method to be easy to measure acoustic wave velocities suitable for the above mentioned disk samples by an ultrasonic thickness gauge with high frequency pulse oscillation. This method was applied to hard and soft PZT[7] and lead-free ceramics composed of alkali niobate[8,9] and alkali bismuth titanate[10]. In this pursuit, we report the fluctuations of acoustic wave velocities in PZT ceramics by this method and the calculation results of Young's modulus and Poisson's ratio, especially to obtain high piezoelectricity in lead-free ceramics.

## EXPERIMEꞨTAL PROCEDURE

An ultrasonic precision thickness gauge (Olympus Co., Model 35DL) used has PZT transducers with 30 MHz for longitudinal wave generation and 20 MHz for transverse wave generation. The acoustic wave velocities were evaluated by the propagation time between second pulse echoes in thickness of ceramic disks with dimensions of 14 mm in diameter and 0.5-1.5 mm in thickness. The sample thickness was measured by a precision micrometer (Mitutoyo Co., Model MDE-25PJ). Piezoelectric ceramic compositions measured were as follows: $0.05Pb(Sn_{0.5}Sb_{0.5})O_3$-$(0.95-x)PbTiO_3$-$xPbZrO_3$ (x=0.33, 0.45, 0.48, 0.66, 0.75) with (hard PZT) and without 0.4 wt% $MnO_2$ (soft PZT)[7]; $(1-x)(Ꞧa,K,Li,Ba)(Ꞧb_{0.9}Ta_{0.1})O_3$-$xSrZrO_3$(SZ) (x=0, 0.02, 0.04, 0.05, 0.06, 0.07)[8,9]; $(1-x)(Ꞧa_{0.5}Bi_{0.5})TiO_3(ꞨBT)$-$x(K_{0.5}Bi_{0.5})TiO_3(KBT)$ (x=0.08, 0.11, 0.15, 0.18, 0.21, 0.28) and $0.79ꞨBT$-$0.20KBT$-$0.01Bi(Fe_{0.5}Ti_{10.5})O_3$ (BFT)[10]; and $(1-x)ꞨBT$-$xBaTiO_3$(BT) (x=0.03, 0.07, 0.11)[10]. The ceramics were prepared through the processes of normal firing in air atmosphere, oxygen atmosphere firing to realize pore-free ceramics[11], as-fired (before poling) and after poling.

## RESULTS AꞨD DISCUSSIOꞨ
### Fluctuation of Longitudinal and Transverse Waves by Pulse Echo Measurement

Figure 1 shows pulse echoes of longitudinal acoustic wave in hard PZT ceramics at a composition of x=0.48. The longitudinal wave velocity was calculated from the propagation time between second pulse echoes ( ⌣ ) and the thickness of ceramic disks. The dependences of the longitudinal and transverse wave velocities on the composition of x in hard and soft PZT ceramics and their fluctuation were shown in Figs. 2(a)-(d). The fluctuation of the velocities in transverse wave was smaller than the ones in longitudinal wave when the samples of n=16-21 pieces were measured at each composition. In addition, it was clarified that the fluctuation in soft PZT was smaller than the one in hard PZT ceramics because of easy movement of ferroelectric domains by poling field, namely low coercive field.

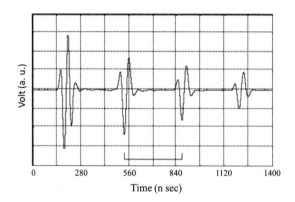

Figure 1. Pulse echoes of longitudinal acoustic wave in hard PZT ceramics (disk dimensions: 13.66 mm diameter and 0.735 mm thickness) before poling at x=0.48; the calculated wave velocity is 4,319 m/s.

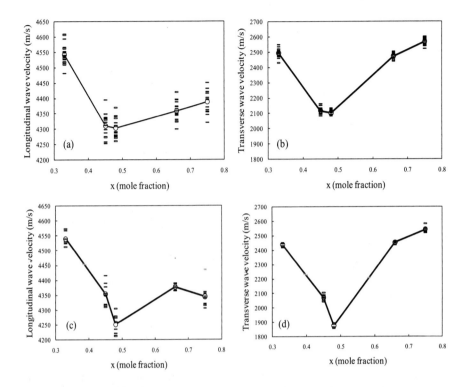

Figure 2. Composition x dependence of longitudinal and transverse wave velocities in hard [(a) and (b)] and soft PZT [(c) and (d)] ceramics before poling; all the samples of n=16-21 are shown in the figure to evaluate their fluctuation.

Young's Modulus and Poisson's Ratio in PZT Ceramics

Figures 3(a)-(d) show the dependences of the longitudinal and transverse wave velocities, Young's modulus and Poisson's ratio on the composition of x in hard and soft PZT ceramics, respectively. The large differences in these values between hard and soft PZT ceramics only occurred around a composition of x=0.48, which corresponds to a morphtropic phase boundary (MPB)[11]. The values of pore-free ceramics fired in oxygen atmosphere also show in the figures at compositions of x=0.66 and 0.75 (in dotted circles, the samples of n=2). The effect of reducing pores in ceramics on the values works like the increase of the longitudinal wave velocity and Poisson's ratio, and the decrease of the transverse wave velocity and Young's modulus. Therefore, it was confirmed by using the pulse echo measurement that oxygen atmosphere firing is an effective tool to improve piezoelectric properties[11], which correspond to decrease Young's modulus and increase Poisson's ratio, as well as production of pore-free ceramics.

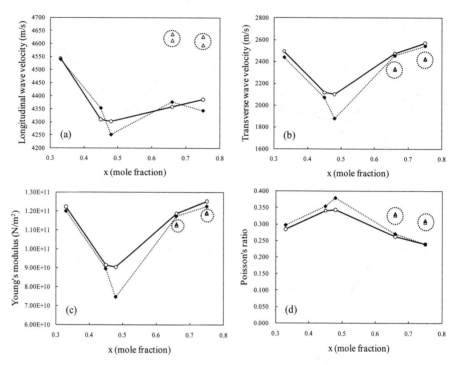

Figure 3. Composition x dependence of (a) longitudinal and (b) transverse wave velocities, (c) Young's modulus and (d) Poisson's ratio in hard (—) and soft (---) PZT ceramics before poling; the values of pore-free hard PZT ceramics (x=0.66 and 0.75, the samples of n=2) were shown in dotted circles.

Figures 4(a)-(d) show the dependences of the longitudinal and transverse wave velocities, Young's modulus and Poisson's ratio on the firing position in sagger at a composition of x=0.45 in hard PZT ceramics, respectively. Nos. 1 and 10 in the figure correspond to the top position piled up and the bottom position piled up in sagger (Fig. 5). It became clear that there was fluctuation of the velocities and elastic constants in the case of as-fired (before poling) samples. In addition, the sample of the middle position piled up of No. 5 possesses low wave velocities, low Young's modulus and high Poisson's ratio because of the firing under high lead oxide (PbO) atmosphere. On the other hand, the samples of the top and bottom positions of Nos. 1, 2 and 10 possess high wave velocities, high Young's modulus and low Poisson's ratio because of the firing under low PbO atmosphere due to vaporization of PbO during firing[11]. Although Young's modulus and Poisson's ratio in PZT ceramics of No. 5 could be improved by high PbO atmosphere firing as shown in Figs. 4(c) and (d), there is a different tendency to decrease the longitudinal wave velocity in comparison with the case of oxygen atmosphere firing, which also can be improved the elastic constants. It is thought that the fluctuation of as-fired ceramics connected to the fluctuation of ceramics after poling.

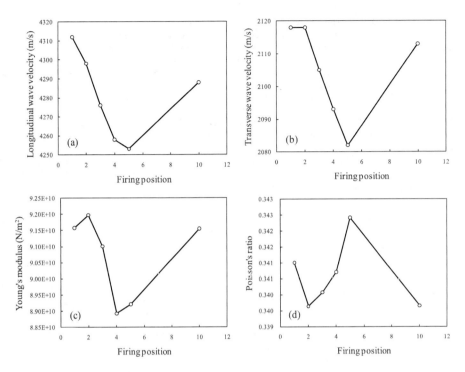

Figure 4. Firing position (Nos. 1 and 10 correspond to ceramic disks of top piled up and bottom piled up in sagger) dependence of (a) longitudinal and (b) transverse wave velocities, (c) Young's modulus and (d) Poisson's ratio in hard PZT ceramics at x=0.45.

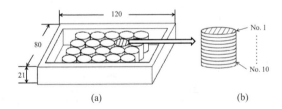

Figure 5. Schematic pictures of (a) PZT ceramic disks piled up in dense Al$_2$O$_3$ sagger (container dimensions: 120 mm length, 80 mm width and 21 mm height) and (b) firing positions (Nos. 1-10) of ceramic disks (14 mm diameter and 1.0 mm thickness); the ceramic disk of No. 5 is surrounded with high PbO atmosphere by the other disks piled up, and it is understood that PbO vaporization is easy to occur at the disks of Nos.1, 2 (top) No.10 (bottom) even through a cover (the same of the container) is put on the container while firing[11].

Young's Modulus and Poisson's Ratio in Lead-Free Ceramics

Figures 6(a)-(d) show the dependences of the longitudinal and transverse wave velocities, Young's modulus and Poisson's ratio on the composition of lead-free ceramics after poling; the compositions were $(1-x)(Na,K,Li,Ba)(Nb_{0.9}Ta_{0.1})O_3$-xSZ (x=0, 0.02, 0.04, 0.05, 0.06, 0.07), $(1-x)$NBT-xKBT (x=0.08, 0.11, 0.15, 0.18, 0.21, 0.28), 0.79NBT-0.20KBT-0.01BFT and $(1-x)$NBT-xBT (x=0.03, 0.07, 0.11), respectively. While the compositions to obtain high electro-mechanical coupling factor in $(Na,K,Li,Ba)(Nb_{0.9}Ta_{0.1})O_3$-SZ (x=0.05-0.06) and NBT-KBT (x=0.18)[9,10] correspond to the compositions with low Young's modulus and high Poisson's ratio as well as PZT (x=0.48). Although a MPB existed around x=0.18 in the phase diagram of NBT-KBT[12,13], a MPB did not exist in the phase diagram of $(Na,K,Li,Ba)(Nb_{0.9}Ta_{0.1})O_3$-SZ. On the other hand, the compositions with high coupling factor in NBT-BT (x=0.07)[10] possessed low Young's modulus and low Poisson's ratio such as PbTiO$_3$[14]. In the phase diagram of NBT-BT, it is confirmed that there is a MPB near x=0.07[15]. Furthermore, there was no significant difference in Young's modulus and Poisson's ratio between as-fired ($\triangle$) (before poling) and after poling ($\blacktriangle$) in $(Na,K,Li,Ba)(Nb_{0.9}Ta_{0.1})O_3$-SZ.

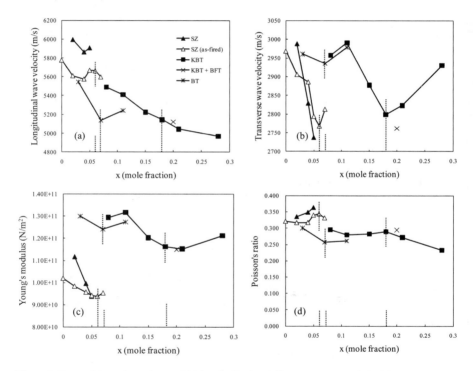

Figure 6. Composition x dependence of (a) longitudinal and (b) transverse wave velocities, (c) Young's modulus and (d) Poisson's ratio in lead-free ceramics of $(1-x)(Na,K,Li,Ba)(Nb_{0.9}Ta_{0.1})O_3$-xSZ (x=0, 0.02, 0.04, 0.05, 0.06, 0.07), $(1-x)$NBT-xKBT (x=0.08, 0.11, 0.15, 0.18, 0.21, 0.28), 0.79NBT-0.20KBT-0.01BFT and $(1-x)$NBT-xBT (x=0.03, 0.07, 0.11).

From the above mentioned viewpoint of the study on the elastic constants in lead-free ceramic, it was clarified that there are compositions with high coupling factor in the cases of (1) low Young's modulus and high Poisson's ratio (PZT type) and (2) low Young's modulus and low Poisson's ratio (PbTiO₃ type). Namely, it was found that there are two kinds of MPB compositions, (1) and (2) in lead-free ceramics with high coupling factor. Therefore, lead-free ceramic compositions with high piezoelectricity must be primarily focused on to realize the compositions with low Young's modulus such as MPB compositions[16]. Secondarily, the importance to measure Poisson's ratio was understood for the research to recognize the compositions of (1) and (2) in lead-free ceramics with high piezoelectricity.

## CONCLUSIONS

Longitudinal and transverse wave velocities of PZT and lead-free ceramics were measured by an ultrasonic precision thickness gauge with high frequency pulse oscillation to calculate elastic constants such as Young's modulus and Poisson's ratio. Although there was the fluctuation of the wave velocities, it was possible to evaluate the elastic constants. The dependences of compositions and firing processes of the ceramics on the elastic constants were clarified in PZT ceramics. Moreover, while the compositions around MPB in lead-free ceramics were investigated by this method, it was found significant relationships between Young's modulus and Poisson's ratio in lead-free ceramics with high piezoelectricity.

## ACKNOWLEDGEMENTS

This work was partially supported by a Grant-in-Aid for Scientific Research C (No. 21560340) and a Grant of Strategic Research Foundation Grant-aided Project for Private Universities (No. S1001032) from the Ministry of Education, Culture, Sports, Science and Technology.

## REFERENCES

[1]M. Kadota, Y. Saito, T. Yoshida, H. Ozeki, T. Yamaguchi, M. Hasegawa, and H. Sato, *Piezoelectric Ceramics Technical Committee, EMAS-6100, Standard of Electronic Materials Manufacturers Association of Japan*, (1993) [in Japanese].
[2]R. Bechmann et al., IRE Standards on Piezoelectric Crystals, *Proc. IRE*, 1958, p. 764.
[3]K. Shibayama et al., *Danseiha Soshi Gijyutsu Handbook* (Technical Handbook of Acoustic Wave Devices) (Ohmsha, Tokyo, 1991), p. 29 [in Japanese].
[4]W. P. Mason et al., *Physical Acoustics I, Part A*, (Academic Press, New York, 1964), p.182.
[5]K. Takagi et al., *Cyouonpa Binran* (Ultrasonic Handbook) (Maruzen, Tokyo, 1999) p. 395 [in Japanese].
[6]K. Negishi and K. Takagi, *Cyouonpa Gijutsu* (Ultrasonic Technology) (Univ. of Tokyo Press, Tokyo, 1984) p.109 [in Japanese].
[7]T. Ogawa and K. Nakamura, Effect of Domain Switching and Rotation on Dielectric and Piezoelectric Properties in Lead Zirconate Titanate, *Jpn. J. Appl. Phys.*, **38**, 5465-5469 (1999).
[8]M. Furukawa T. Tsukada, D. Tanaka, and N. Sakamoto, Alkaline Niobate-based Lead-Free Piezoelectric Ceramics, *Proc. 24th Int. Japan-Korea Semin. Ceramics*, 2007, p. 339-342.
[9]T. Ogawa, M. Furukawa, and T. Tsukada, Poling Field Dependence of Piezoelectric Properties and Hysteresis Loops of Polarization versus Electric Field in Alkali Niobate Ceramics, *Jpn. J. Appl. Phys.*, **48**, 709KD07 (2009).
[10]T. Ogawa, T. Nishina, M. Furukawa, and T. Tsukada, Poling Field Dependence of Ferroelectric Properties in Alkali Bismuth Titanate Lead-Free Ceramics, *Jpn. J. Appl. Phys.*, **49**, 09MD07 (2010).
[11]T. Ogawa, Highly Functional and High-Performance Piezoelectric Ceramics, *Ceramic Bulletin*, **70**, 1042-1049 (1991).

[12]W. Zhao, H. P. Zhou, Y. K. Yan, and D. Liu, *Key Eng. Mater.*, **368-372**, 1908 (2008).

[13]Z. P. Yang, B. Liu, L. L. Wei, and Y. T. Hou, *Mater. Res. Bull.*, **43**, 81 (2008).

[14]T. Ogawa, Poling Field Dependence of Crystal Orientation and Ferroelectric Properties in Lead Titanate Ceramics, *Jpn. J. Appl. Phys.*, **39**, 5538-5541 (2000).

[15]Y. J. Dai, J. S. Pan, and X. W. Zhang, *Key Eng. Mater.*, **336-338**, 206 (2007).

[16]Y. Saito, H. Takao, T. Tani, T. Ionoyama, K. Ta katori, T. Homma, T. Iagaya, and M. Iakamura, *Iature* , **432**, 84 (2004).

# LONG-TERM AND LIGHT STIMULATED EVOLUTION OF SEMICONDUCTOR PROPERTIES

Sergei Pyshkin
Institute of Applied Physics, Academy of Sciences
Kishinev, Moldova

John Ballato, George Chumanov, and Donald VanDerveer
Clemson University, Clemson, SC, USA

Raisa Zhitaru
Institute of Applied Physics, Academy of Sciences
Kishinev, Moldova

## ABSTRACT

We demonstrate that long-term natural stimuli can dramatically improve the perfection of semiconductor crystals grown under laboratory conditions. More specifically, diffusion and stress relaxation lead over time to host atoms migrating into their proper equilibrium atomic positions while impurities uniformly redistribution. We demonstrate that the highly ordered crystallinity that develops over several decades enhances stimulated emission, increases the radiative recombination efficiency and spectral range of luminescence. For instance, at room temperature as in conventional GaP nanoparticles, the bulk long-term ordered single crystals, having the forbidden gap around 2.2 eV, develop a bright luminescence until 3.2 eV.

Along with the long-term ordering, the photomechanical effect plays an important role in the formation of close to ideal crystals. Investigating behavior of Raman light scattering, X-ray diffraction, microhardness and photoluminescence properties of GaP single crystals, we note that this effect is not very essential for freshly prepared GaP crystals, but it acts as an additional shaking of the close to ideal crystal lattice or powerful triggering mechanism which finally improves mechanical and optical characteristics of the crystal after its 50 years long-term ordering at normal pressure and room temperature.

Thus, dramatically increasing the efficiency of this one of the basic materials for micro- and nanoelectronic industry, we considerably improve opportunities for its wide and interesting application.

## INTRODUCTION

Single crystals of semiconductors grown under laboratory conditions naturally contain a varied assortment of defects such as displaced host and impurity atoms, vacancies, dislocations, and impurity clusters. These defects result from the relatively rapid growth conditions and inevitably lead to the deterioration of the mechanical, electric, and optical properties of the material, and therefore to degradation in the performance of the associated devices.

Over time, driving forces such as diffusion along concentration gradients, strain relaxation associated with clustering, and minimization of the free energy associated with properly directed chemical bonds between host atoms result in ordered redistribution of impurities and host atoms in a crystal.

The pure and doped GaP crystals discussed herein were prepared app. 50 years ago. Investigating their electro- and photoluminescence (PL), photoconductivity, bound excitons, nonlinear optics, and other phenomena, it was of interest also to monitor the change in crystal quality over the course of several decades while the crystal is held under ambient conditions.

We demonstrate that along with the long-term ordering the photomechanical effect [1] plays an important role in the formation of close to ideal crystals and it was also applied at the final stage of these experiments.

GaP is one of the key light emissive materials for development of optoelectronics and photonics. One of the possible ways to increase its luminescence characteristics is improvement of its crystal lattice to the state when the lattice defects do not influence on the efficiency of irradiative electron-hole recombination in optical transitions equally between the minimum and maximum of the conductance and the valence bands as well as between higher and lower specific points of its Brillouin zone.

Obviously, dramatically increasing the efficiency of this basic material and applying only one compound and technology of its preparation and processing instead of a lot of them, we considerably improve situation in micro- and nanoelectronic industry.

METHODS OF CHARACTERIZATION

Earlier it was shown that the successful thermal processing of GaP improving quality of the crystal lattice and ordering of impurity disposition can only take place at temperatures below its sublimation temperature, requiring a longer annealing time. Evaluated in framework of the Ising model the characteristic time of the substitution reaction during N diffusion along P sites in GaP:N crystals at room temperature constitutes 15-20 years [2]. Hence, the observations of some properties of the crystals made in the sixties and the nineties were then compared with the results obtained in 2009-2010 in closed experimental conditions.

Comparison of the properties of the same crystals has been performed in the 1960s, 1970s, 1980s, 1990s [2-10], and during 2000s [11–33] along with those of newly made GaP nanocrystals [15, 27-29, 31, 32] and freshly prepared bulk single crystals [21-25].

The influence of long-term ordering on optical and mechanical properties has been studied using measurements of photoluminescence (PL), photomechanical effect (PME), microhardness, dislocation density, Raman light scattering (RLS) and X-ray diffraction (XRD). Combination of these characterization techniques elucidates the evolution of these crystals over the course of many years, the ordered state brought about by prolonged room-temperature thermal annealing, and the interesting optical properties that accompany such ordering.

The influence of long-term on microhardness of GaP crystals has been studied using a PMT-3 microscope equipped by Vickers pyramid indenter. The load to the indenter was 50g. Deformation zones around the impressions of the indenter were investigated at room temperature by a near-field scanning optical microscope.

Raman spectra were obtained using an Innova 200 $Ar^+$ ion laser operated at 514.5 nm and a Triplemate 1377 spectrograph interfaced to the Princeton Instruments Model LN1152a liquid nitrogen-cooled detector. A computerized acquisition system recorded the relative signal detected by the photon counting technique. The measurements were taken at room temperature and in a cold finger configuration at a sample temperature of approximately 80 K. New results and the results spanning the years 1989-1993 were obtained under similar experimental conditions ($Ar^+$ laser at 514.5 nm), using the same spectrographs and acquisition systems.

X-ray diffraction data were collected on Rigaku ULTIMA IV powder diffractometer using a monochromator and Cu Kα radiation (1.5406 Å). All scans were in the θ-θ mode at 300K.

Photomechanical effect was provided using the 337.1 nm, 15 ns, 50Hz Q-switched $N_2$ laser with tunable between 10-360 KWt/cm$^2$ density of power and 1-30 min exposition time. The same laser was used for excitation of luminescence.

In spite of the long-term character of the measurements the same experimental conditions were strictly adhered. In addition, freshly prepared GaP crystals were purchased for comparison with the samples from our unique collection of the long-term ordered GaP crystals. The study of freshly-prepared crystals provides a good opportunity to compare these long-term measurements obtained on the old equipment with freshly grown crystals.

RESULTS AND DISCUSSION

This work continues monitoring of properties of GaP single crystals grown in 1960s [3] and for years investigated by us [2-35, 37]. With the recent achievements of nanotechnology we get a unique opportunity to compare properties of the long-term ordered bulk GaP bulk single crystals with synthesized by us GaP nanocrystals and GaP/polymers nanocomposites [15, 26-33].

It was earlier shown [10-27, 33], the long-ordering of N impurity and host atoms positions in crystal lattice of GaP leads to creation of a new type of the crystal lattice where the host atoms occupy their proper (equilibrium) positions in the crystal field, while the impurities, once periodically inserted into the lattice, divide it in the short chains of equal length, where the host atoms develop harmonic vibrations. This periodic substitution of a host atom by an impurity allows the impurity to participate in the formation of the crystal's energy bands. It leads to the change in the value of the forbidden energy gap, to improvement of mechanical and optical properties, to the appearance of the discussed below phenomena, interesting for study and application in optoelectronics.

Let us discuss the state of the long-term ordered GaP:N crystals before the undertaken recently light stimulated evolution of their mechanical and optical properties.

Taken from [27] Figure 1 shows the absorption edge as a function of photon energy (Fig.1a) and provides a comparison of microhardness and density of dislocations as a function of N concentration (Fig. 1b). The position of the absorption edge in freshly prepared GaP:N crystals does not depend on N impurity concentration and it coincides with its position in pure GaP crystals (Fig. 1a, curves 2 and 3). This N impurity creates the 21-meV energy level under the conduction band for bound excitons both in freshly prepared and 25-year ordered crystals. On the contrary, curve 1, Fig. 1a shows that the GaP:N crystals ordered for approximately 40 years and more demonstrate an increase in the forbidden gap and the clear shift of the absorption edge, which is expected to be proportional to the N concentration according to the Vegard law, similar to that in a dilute GaP-GaN solid solution.

Long-term ordering of the dopants and defects changes the mechanical properties of GaP and its ternary analog $CdIn_2S_4$.

It is known [38], good plasticity is determined by free movement of dislocations through the crystal under a mechanical load. Impurities act to pin the movement of dislocations and due to this fact the value of microhardness, H, in GaP depends on the impurity concentration (Fig.1b, lines 1 and 2). One can see, the relatively pure crystals have minimum microhardness. An increase in the impurity concentration in GaP crystals leads to an increase in microhardness for the long-term-ordered crystals (Fig.1b, lines 1 and 2, respectively).

A rather large difference in microhardness of the long-term-ordered highly doped crystal relative to the newly grown crystals is explained by the regular disposition of impurities. This regular disposition of impurities creates a more significant obstacle to dislocation movement than in the newly grown system, in which the impurities form clusters with large distances between them, permitting greater dislocation movement. Indeed, at the concentration of about $N = 10^{18} cm^{-3}$ with the ordered disposition of the impurities in GaP crystal lattice, the 10 nm distance between the impurities is a serious obstacle for dislocation movement through this regular impurity net taking into account the dislocation density of the order of $10^4 cm^{-2}$ (please see Fig.1) and the significant lateral extent of the dislocations. However, at the same concentrations of impurities and dislocation density, the distances between impurity clusters in disordered crystal can considerably exceed 10 nm by a few orders of magnitude, resulting in a rather free dislocation movement and a corresponding decrease in microhardness.

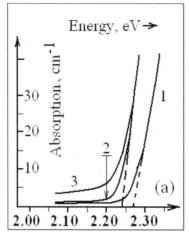 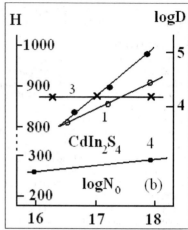

Figure 1. a. Absorption coefficient in long-term ordered (40 years and more) GaP:N (1) and in 25 year old (2) or freshly prepared undoped GaP (3). N concentration is $10^{18}$ cm$^{-3}$.
b. Microhardness H (Kg/mm$^2$) in fresh unordered GaP crystals as a function of N concentration (1); the same for 40 years and more ordered crystals (2); Density of dislocations (cm$^{-2}$) D as a function of N concentration (3); H in CdIn$_2$S$_4$ at different concentration of defects (4).

As noted previously, the luminescence of fresh doped and undoped crystals could be observed only at temperatures below about 80 K. The luminescence band and lines were always seen at photon energies less than the value of the forbidden gap (2.3 eV). Now, after the decades from 1960s, luminescence of the long-term-ordered bulk crystals similar to the GaP nanocrystals [28, 29, 31-33] is clearly detected in the region from 2.0 eV to 3.4 eV at room temperature [13-27].

Note that increase of luminescence excitation in the case of partly ordered GaP:N (Fig. 2a) leads to a broad luminescence band as a result of bound exciton interaction [8], while in the case of perfectly ordered crystals (Fig. 2b) one can see an abrupt narrowing of the luminescence band, probably, due to stimulated emission in defect-free crystals. The density of power of the pumping nanosecond laser at 50 mJ energy in the pulse is usually enough for stimulated emission in a perfect GaP single crystal with natural faceting.

Earlier, in freshly prepared crystals we observed a clear stimulated emission from a GaP:N resonator at 80 K [34] as well as so called superluminescence from the GaP single crystals having natural faceting. Presently, our ordered crystals have a bright luminescence at room temperature that implies their perfection and very lower light losses. The narrowing of the luminescence band in Fig. 2b can probably be explained by stimulated emission at its initial stage of superluminescence. In our current presentations and papers [23, 24, 27] we demonstrate that the stimulated emission is developed even at room temperature by direct electron–hole recombination of an electron at the bottom of the conduction band with a hole at the top of the valence band and the LO phonon absorption.

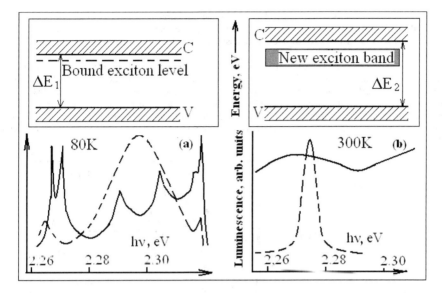

Figure 2. Luminescence spectra and the view of the forbidden gaps ($\Delta E_1$, $\Delta E_2$) in the partly 25 year ordered (a) and perfect 40 years and more ordered (b) crystals GaP doped by N. The dotted lines correspond to highly optically excited crystals. C and V – the positions of the bottom of the conductance and the top of valence bands respectively.

Thus, the recent experiments on photomechanical effect were carried out on the long-term ordered, close to ideal GaP:N single crystals grown in 1960s. The rather perfect state of their crystal lattice and impurity superlattice is confirmed by many details of their semicentennial evolution in normal ambient environments obtained through investigations of the same crystals since the moment of their growth until present by measurements of microhardness and by the methods of Raman light scattering (RLS), X-ray diffraction and photoluminescence (PL), described in [2-33]. The results of this lasting for decades evolution have been compared with the changes in same crystals after very short (20-30 min) light irradiation by the laser with noted above parameters. Let us discuss some of these changes.

For the beginning we note that the mechanical properties of aged app. 50 years GaP:N crystals (microhardness, fragility and plasticity, the general view of irradiated crystal surface) are changing for a long time after laser irradiation (a month and more) and as the result their hardness increases in proportion to the N impurity concentration. For instance, during 20 days after laser irradiation microhardness of heavy doped GaP:N crystals changes in the limits 800 - 1000 kg/mm$^2$ and more, whereas undoped long-term ordered GaP single crystals demonstrate some decrease of their microhardness.

As it was shown in Fig.1b, the pure long-term ordered GaP crystals have minimum microhardness that confirms their plasticity as the result of relatively free movement of dislocations. And additional laser irradiation stimulates the host atoms to occupy their equilibrium positions that improves their plasticity and decreases microhardness.

A rather large difference in microhardness of the laser irradiated and non-irradiated long-term-ordered highly doped crystal is explained by further improvement of impurity superlattice quality in the irradiated crystals that creates additional obstacle to dislocation movement and increases their microhardness.

Obviously, as the result of laser irradiation and the photomechanical effect we have even more perfect crystal quality then in the case of their long-term ordering without the conclusive laser irradiation. This effect, insufficient for freshly prepared GaP crystals, acts as an additional shaking of the close to ideal crystal lattice or powerful triggering mechanism which finally improves mechanical and optical characteristics of the crystal on the background of its 50 years long-term ordering at normal pressure and room temperature. The perfect crystal quality is reflected in shown below spectra of RLS, XRD and PL.

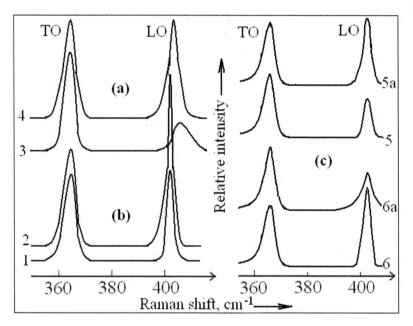

Figure 3. Raman light scattering in undoped GaP and GaP:N in 1989-1993 (a) and in 2006 (b) [4, 5, 11, 18, 19] in comparison in with the same, but laser irradiated crystals (c) .
1, 4, 5a – pure non-irradiated GaP; 2,3, 6a – GaP heavy doped by N non-irradiated crystals; 5, 6 – the same, long-term ordered (app. 50 years) but laser irradiated pure (5) and doped (6) crystals.

The difference in the present ordered state of the laser irradiated crystal lattice, with respect to the data obtained with the same, but not laser irradiated crystals and conditions in 2006 [27] and during 1989–1993 [10, 11] can be seen in Fig.3a, b and c, which shows the Raman spectra of pure GaP and GaP:N [18, 19]. One can see (Fig.3a, curves 3 and 4) that for the longitudinal optical (LO) phonon modes the peak from the original samples is broad, weak, and shifts with impurity concentration. After 40 years (Fig.3a, curves 1 and 2), the peak for heavily N-doped GaP is much more intense, has a more symmetric (Lorentzian) shape, narrower linewidth, and a spectral position that no longer depends on N concentration. These results are characteristic of harmonic vibrations in a more perfect lattice.

The LO phonon line is narrower in the aged doped crystal (Fig.3a, curve 2) than in the undoped (pure) crystals (Fig.3a, curve 1) and is also more intense than the TO phonon line. Similar results have been obtained for various impurities (N, Sm, and Bi) in spite of their maximum possible concentrations in GaP and different masses of impurity atoms and different types of substitution [18]. These results confirm the important role of the impurities that are periodically located in the host crystal lattice and result in the formation of the new perfect crystal lattice. Note that the RLS spectra are nearly the same (only slightly narrowing of the LO lines is observed for irradiated samples) for pure initial and laser irradiated GaP crystals (Fig.3c, spectra 5a and 5), but doped by nitrogen GaP:N crystals (Fig.3c, spectra 6a and 6) after laser irradiation demonstrate additional improvement of their quality by narrowing of very symmetrical Lorentzian shape of LO phonon lines and, just opposite to the situation in pure aged GaP crystals, considerable redistribution of TO and LO phonon concentrations in favor of LO phonons in short and perfect chains of crystal lattice between periodically located isoelectronic N impurities substituting host P atoms. Thus, comparing RLS spectra of the long-term ordered GaP before and after irradiation, we note the further improvement of its crystal lattice and it is especially distinctly in the case of N doped single crystals.

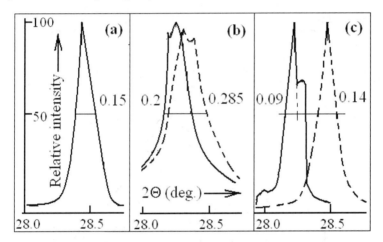

Figure 4. X-ray diffraction from grown in 2007 (a) and long-term ordered non-irradiated (dotted lines) and laser irradiated (solid lines) GaP single crystals grown in 1960s. Only one curve in Fig. 4a corresponds to irradiated and non-irradiated crystals. Halfwidths are shown on the left (laser irradiated crystal) and on the right (non-irradiated) from the relative spectra.

XRD spectra of freshly prepared and long-term ordered non-irradiated and irradiated GaP:N crystals shown in Fig.4 are in a good agreement with the results obtained from their characterization by the methods of microhardness and RLS. One can see that XRD spectra of freshly prepared GaP crystals (kindly supplied by Aldrich Co.) are the same for irradiated and non-irradiated crystals (Fig. 4a), whereas the characteristic for GaP lines in the $\Theta - 2\Theta$ rocking curves at 28 degree show some narrowing for pure ordered crystals and a considerable narrowing and additional fine structure for N doped ordered crystals.

At last, using GaP:N as an example let us discuss present and available in principle results in luminescence of a semiconductor if his crystal lattice and distribution of necessary impurities inside the lattice are closed to ideal.

As we noted [27], due to a significant number of defects and a highly intensive non-radiative recombination of non-equilibrium current carriers, initially luminescence of fresh undoped crystals could be observed only at the temperatures 80K and below as it is shown in Fig.2a, but now it is clearly detected in the region from 1.8 eV and until 3.2 eV at room temperature [19, 24, 33]. Taking into account that the indirect forbidden gap is only 2.25 eV, it is suggested that this considerable extension of the region of luminescence to the high energy side of the spectrum as well as a pronounced increase of its brightness are connected with a very small concentration of defects, considerable improvement of crystal lattice and more high transparency of perfect crystals. These reasons create an opportunity to see intense luminescence in optical transitions from higher extrema of the conductance band to the top of the valence band.

Fig.5, spectra 1 and 2, shows the luminescence of close to ideal N doped GaP single crystals grown in special conditions [3] in 1960s and for 20-30 min. laser irradiated after aging during app. 50 years. Spectrum 1 is obtained at 300K from optically excited ($N_2$ pulse laser, 355nm) GaP:N crystal as well as the narrow spectrum 2 is obtained from the same but more highly optically excited crystal. In this conditions we observe at 300K at relatively low level of optical excitation the broad band intense luminescence in all visible region with clear maxima close to indirect $X_1^c$ - $\Gamma_{15}^v$ and direct $\Gamma_1^c$ - $\Gamma_{15}^v$ extrema in the Brillouin zone of GaP respectively at 2.2 and 2.8 eV [36, 37], whereas some increase of excitation level leads to sharp narrowing of the spectrum due to superluminescence at the indirect transition $X_1^c$ - $\Gamma_{15}^v$ with LO phonon absorption. Note that only ordered and laser irradiated GaP crystals have the very broad luminescence band covering the all visible region at relatively low level of optical excitation, as well as the low threshold of excitation for the appearance of superluminescence.

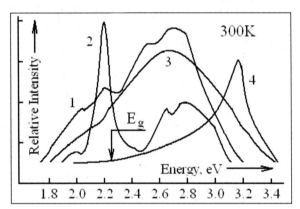

Figure 5. Luminescence of 50 years aged and then 30min laser irradiated bulk GaP single crystals at relatively low (1) and high (2) level of optical excitation in comparison with the luminescence of GaP nanoparticles (3, 4).

Note that the light absorption coefficient in the vicinity of the direct transitions $\Gamma_1^c$ - $\Gamma_{15}^v$, 2.8 eV, and $\Gamma_{15}^c$ - $\Gamma_{15}^v$, 3.2 eV, light emissive transitions exceeds $10^6 cm^{-1}$ [36, 37], therefore, so called "hot" luminescence from high $\Gamma_1^c$ and $\Gamma_{15}^c$ extrema can be seen only in an absolutely perfect GaP crystal with very high quantum efficiency, as it is presented here by the luminescence, Fig.5, spectrum 1, but not in any freshly grown one having a lot of different defects and zero light emissive efficiency at 300K not only from this points but also from minimum extremum $X_1^c$ (2.2 eV) and from the impurity levels inside the forbidden gap. Obviously, an abrupt increase of the absorption coefficient gives an opportunity, changing the thickness of the GaP perfect sample, to modify by a simple way the luminescence spectrum in the necessary for application direction and to change its position of maximum and shape in wide spectral region from yellow-red until ultraviolet (UV).

It is interesting to compare these results obtained from perfect, ordered and laser irradiated bulk GaP single crystals with the best samples of its nanoparticles, obtained by us recently [28, 31]. It is known that nanoparticles are introduced into optoelectronic industry as efficient light emissive matters in which non-radiative recombination processes are suppressed due to the fact that their dimensions are less than the free path of non-equilibrium electron-holes pairs in the material of the particles. Obviously, in this situation there is no need for thermalization of optically excited electrons into the lowest conductance band extremum before their recombination – they can recombine with holes in the valence band directly from the point of the Brillouin zone in which they have been excited and until its minimum point and the resulting photons can not be absorbed inside the crystal due its complete transparency at the thickness of app. 10-50 nm. Spectrum 3 in Fig.5 shows luminescence of non-separated on dimensions nanoparticles closed in its characteristics to the luminescence of aged and laser irradiated bulk GaP crystals, but slightly shifted to UV side due to known quantum confinement effect, whereas the nanoparticles specially separated on the minimum dimensions of app. 10 nm and stored as a suspension in a liquid excited by laser photons with the energy app. 3.5 eV demonstrate bright blue-UV luminescence with the single maximum at 3.2 eV at the vicinity to the nearest $\Gamma_{15}^c$ - $\Gamma_{15}^v$ transition, 3.2 eV (Fig.5, spectrum 4).

CONCLUSIONS

This study together with cited here our works published earlier brings a novel perspective to improvement of the quality of semiconductor crystals. The unique collection of pure and doped crystals of semiconductors grown in the 1960s provides an opportunity to observe the long term evolution of the properties of these key electronic materials and compare the close to ideal (aged and improved through laser irradiation) bulk crystals with their high quality nanoparticles. During this almost half-centennial systematic investigation we have established the main trends of the evolution of optoelectronic and mechanical properties. It was shown that these stimuli to improve quality of the crystal lattice are the consequence of thermodynamic driving forces and prevail over tendencies that would favor disorder.

For the first time, as the results of long-term ordering and additional laser irradiation at the end of this semicentennial process, we observe a new type of the crystal lattice, in which the host atoms occupy their proper (equilibrium) positions in the crystal field, while the impurities, once periodically inserted into the lattice, divide it in the short chains of equal length, where the host atoms develop harmonic vibrations. This periodic substitution of a host atom by an impurity allows the impurity to participate in the formation of the crystal's energy bands. It leads to the change in the value of the forbidden energy gap, to the appearance of a crystalline excitonic phase [8, 11], and to the broad energy bands instead of the energy levels of bound excitons. The high perfection of this new lattice leads to the abrupt decrease of non-radiative mechanisms of electron-hole recombination, to both the relevant increase of efficiency and spectral range of luminescence from yellow-red until blue-ultraviolet region and to the stimulated emission of light due to its amplification inside the well arranged, defect-free medium of the crystal.

Along with the long-term ordering, the photomechanical effect plays an important role in the formation of close to ideal crystals. Investigating behavior of Raman light scattering, X-ray diffraction, microhardness and photoluminescence properties of GaP single crystals, we note that this effect is not very essential for freshly prepared GaP crystals, but it acts as an additional shaking of the close to ideal crystal lattice or powerful triggering mechanism which finally improves mechanical and optical characteristics of the crystal after its 50 years long-term ordering at normal pressure and room temperature.

This long-term evolution of the important properties of our unique collection of semiconductor single crystals promises a novel approach to the development of a new generation of optoelectronic devices. The combined methods of laser assisted and molecular beam epitaxies [39-41] will be applied to fabrication of device structures with artificial periodicity; together with classic methods of crystal growth, they can be used to realize impurity ordering that would yield new types of nanostructures and enhanced optoelectronic device performance.

Note, that semiconductor nanoparticles were introduced into materials science and engineering mainly that to avoid limitations inherent to freshly grown semiconductors with a lot of different defects. Our long-term ordered and therefore close to ideal crystals in general repeat behavior of the best nanoparticles. These perfect crystals are useful for application in top-quality optoelectronic devices as well as they are a new object for development of fundamentals of solid state physics.

Thus, dramatically increasing the efficiency of this one of the basic materials for micro- and nanoelectronic industry, we considerably improve opportunities for its wide and interesting application. In general, concerning practical and commercial use of the results of this semicentennial study we note that the modern optoelectronic industry is in highly need of a universal material for advanced light emissive device structures covering broad spectral region. Properly prepared GaP bulk and nanocrystals could be a good alternative to a lot of now used different semiconductors.

ACKNOWLEDGEMENTS

The authors are very grateful to the US Dept. of State, Inst. of International Exchange, Washington, DC, The US Air Force Office for Scientific Research, the US Office of Naval Research Global, Civilian R&D Foundation, Arlington, VA, Science & Technology Center in Ukraine, Clemson University, SC, University of Central Florida, FL, Istituto di elettronica dello stato solido, CNR, Rome, Italy, Universita degli studi, Cagliari, Italy, Joffe Physico-Technical Institute, St.Petersburg State Polytechnical University, Russia, Institute of Applied Physics and Academy of Sciences of Moldova for support and attention to this protracted (1961-present time) research.

REFERENCES
1. A. B. Gerasimov, G. D. Chiradze and N. G. Kutivadze, On the physical nature of a photomechanical effect, *Semiconductors*, **35(1)**, 72-76, DOI: 10.1134/1.1340292; Translated from *Fizika i Tekhnika Poluprovodnikov*, 35(1), 70–74 (2001).
2. Pyshkin S L, Radautsan S I et al., Processes of Long-Lasting Ordering in Crystals with a Partly Inverse Spinel Structure, *Sov. Phys. Dokl.*, **35(4)**, 301-304 (1990).
3. N.A. Goryunova, S.L. Pyshkin, A.S. Borshchevskii, S.I. Radautsan, G.A. Kaliujnaya, Yu.I. Maximov, and O.G. Peskov, Influence of Impurities and Crystallisation Conditions on Growth of Platelet GaP Crystals, *Growth of Crystals*, **8**, ed. N.N. Sheftal, New York, 68-72 (1969), Symposium on Crystal Growth at the 7th Int Crystallography Congress (Moscow, July 1966).
4. B.M. Ashkinadze, S.L. Pyshkin, A.I. Bobrysheva, E.V. Vitiu, V.A. Kovarsky, A.V. Lelyakov, S.A. Moskalenko, and S.I. Radautsan, Some Non-linear Optical Effects in GaP, *Proceedings of the IXth International Conference on the Physics of Semiconductors* (Moscow, July 23–29, 1968), **2**, 1189-1193 (1968).
5. S. Pyshkin, S. Radautsan, and V. Zenchenko, Raman Spectra of Cd-In-S with Different Cation-Sublattice Ordering, *Sov. Phys. Dokl.* **35 (12)**, 1064-67 (1990).

6. S.L. Pyshkin and A. Anedda, Preparation and Properties of GaP Doped by Rare-Earth Elements, *Proceedings of the 1993 Materials Research Society (MRS) Spring Meeting*, Symposium E, **301,**192-197 (1993).

7. S.L. Pyshkin and A. Anedda, Time-Dependent Behavior of Antistructural Defects and Impurities in Cd-In-S and GaP, ICTMC-XI (Salford, UK, 1997), *Institute of Physics Conference Series, Ternary and Multinary Compounds*, **152**, Section E, 785-89 (1998).

8. S. Pyshkin and L. Zifudin, Excitons in Highly Optically Excited Gallium Phosphide, *J. Lumin.*, **9**, 302-308 (1974).

9. S. Pyshkin, Luminescence of GaP:N:Sm Crystals, *J. Sov. Phys. Semicond.*, **8**, 912-13 (1975).

10. S.L. Pyshkin, A. Anedda, F. Congiu, and A. Mura, Luminescence of the GaP:N Ordered System, *J. Pure Appl. Opt.*, **2**, 499-502 (1993).

11. S.L. Pyshkin (invited), Luminescence of Long-Time Ordered GaP:N, The 103rd ACerS Annual Meeting (Indianapolis, 2001), *ACerS Transaction series*, **126**, 3-10 (2002).

12. Sergei L. Pyshkin, Bound Excitons in Long-Time Ordered GaP:N, *Moldavian Journal of the Physical Sciences*, **1(3)**, 14-19 (2002).

13. Sergei Pyshkin, John Ballato, Advanced Light Emissive Composite Materials for Integrated Optics, Symposium: The Physics and Materials Challenges for Integrated Optics - A Step in the Future for Photonic Devices, *Proc of the 2005 MS&T Conference*, Pittsburgh, TMS, 3-13 (2005)

14. J. Ballato and S.L. Pyshkin, Advanced Light Emissive Materials for Novel Optical Displays, Lasers, Waveguides, and Amplifiers, *Moldavian J. of Physical Sciences, 5(2)*, 195-208 *(2006)*.

15. S.L. Pyshkin, J. Ballato, G. Chumanov, J. DiMaio, and A.K. Saha, Preparation and Characterization of Nanocrystalline GaP, *Technical Proceedings of the 2006 ISTI Ianotech Conference*, **3**, 194-197 (2006).

16. S.L. Pyshkin, R.P. Zhitaru, and J. Ballato, Long-Term Evolution of Optical and Mechanical Properties in Gallium Phosphide, *Proceedings of the XVII St. Petersburg Readings on the Problems of Durability*, Devoted to the 90th Birthday of Prof. A.N. Orlov, **2,** 174-176 (2007).

17. S.L. Pyshkin, R.P. Zhitaru, and J. Ballato, Modification of Crystal Lattice by Impurity Ordering in GaP, Int. Symposium on Defects, *Proceedings of the MS&T 2007 Conference*, International Symposium on Defects, Transport and Related Phenomena (Detroit, MI, September 16–20, 2007), 303-310 (2007).

18. S.L. Pyshkin, J. Ballato, and G. Chumanov, Raman light scattering from long-term ordered GaP single crystals, *J. Opt. A. Pure Appl. Opt.*, **9**, 33-36 (2007).

19. S.L. Pyshkin, J. Ballato, M. Bass, and G. Turri (invited), Luminescence of Long-Term Ordered Pure and Doped Gallium Phosphide, TMS Annual Meeting, Symposium: Advances in Semiconductor, Electro Optic and Radio Frequency Materials (March 9–13, New Orleans, LA). *J. Electronic Materials,* **37(4)**, 388-395 (2008).

20. S. Pyshkin, and J. Ballato, Long-Term Ordered Crystals and Their Multi-Layered Film Analogues, *Proceedings of the 2008 MS&T Conference*, Pittsburgh Symposium on Fundamentals & Characterization, Session "Recent Advances in Growth of Thin Film Materials", 889-900 (2008).

21. S.L. Pyshkin, J. Ballato, M. Bass, G. Chumanov, and G. Turri, Time-dependent evolution of crystal lattice, defects and impurities in CdIn$_2$S$_4$ and GaP, *Phys. Stat. Sol.*, **C 6**, 1112-15 (2009).

22. S. Pyshkin, R. Zhitaru, J. Ballato, G. Chumanov, and M. Bass, Structural Characterization of Long Term Ordered Semiconductors , *Proceedings of the 2009 MS&T Conference*, International Symposium "Fundamentals & Characterization," 698-709 (2009).

23. S. Pyshkin, J. Ballato, M. Bass, G. Chumanov, and G. Turri, Properties of the Long-term Ordered Semiconductors, *The 2009 TMS Annual Meeting and Exhibition, Suppl. Proc.*, (San Francisco, February 15–19, 2009), **3**, 477-484 (2009).

24. S. Pyshkin, J. Ballato, M. Bass, and G. Turri, Evolution of Luminescence from Doped Gallium Phosphide over 40 Years, *J. Electronic Materials,* **38(5)**, 640-646 (2009).

25. S. Pyshkin, J. Ballato, G. Chumanov, M. Bass, G. Turri, R. Zhitaru, and V. Tazlavan, Optical and Mechanical Properties of Long-Term Ordered Semiconductors, The 4th International Conference on Materials Science and Condensed Matter Physics, Kishinev, Sept 23-26, 2008, *Moldavian Journal of the Physical Sciences*, **8(3-4)**, 287-295 (2009).

26. Sergei Pyshkin, John Ballato, Andrea Mura, Marco Marceddu, Luminescence of the GaP:N Long-Term Ordered Single Crystals, *Suppl. Proceedings of the 2010 TMS Annual Meetings* (Seattle, WA, USA, February, 2010, **3**, 47-54 (2010).

27. Sergei Pyshkin and John Ballato, Evolution of Optical and Mechanical Properties of Semiconductors over 40 Years, *J. Electronic Materials*, Springer, DOI: 10.1007/s11664-010-1170-z, **39(6)**, 635-641 (2010).

28. S. Pyshkin, J. Ballato, G. Chumanov, N. Tsyntsaru, E. Rusu, Preparation and Characterization of Nanocrystalline GaP for Advanced Light Emissive Device Structures, The 2010 Nanotech Conference (Anaheim, CA, June 21-24), *ISTI, ISTI-Ianotech 2010*, *www.nsti.org, ISBI 978-1-4398-3401-5*, **1**, 522-525 (2010).

29. S. Pyshkin, J. Ballato, I. Luzinov, B. Zdyrko (2010) Fabrication and Characterization of the GaP/Polymer Nanocomposites for Advanced Light Emissive Device Structures, The 2010 Nanotech Conference (Anaheim, CA, June 21-24), *ISTI, ISTI-Ianotech 2010, www.nsti.org, ISBI 978-1-4398-3401-5*, **1**, 772-775 (2010).

30. S. Pyshkin (Project Manager), Joint Moldova/US/Italy/France/Romania STCU (www.stcu.int) Project 4610 "Advanced Light Emissive Device Structures", 2009-2012

31. Pyshkin SL, Ballato J, Belevschii S, Rusu E, Racu A, Van DerVeer D, Synthesis and Characterization of GaP Nanoparticles for Light Emissive Devices. The 2011 Nanotech Conference (Boston, MA, June 13-16), NSTI, NSTI-Nanotech 2011, www.nsti.org, ISBN 978-1 4398-7142-3 **1**, 327-330 (2011).

32. Pyshkin S. L., Ballato J., Luzinov I. & Zdyrko B., Fabrication and characterization of GaP/polymer nanocomposites for advanced light emissive device structures, *Journal of Ianoparticle Research* (2011). DOI 10.1007/s11051-011-0547-0

33. Sergei L. Pyshkin and John M. Ballato, Long-Term Convergence of Bulk- and Nano-Crystal Properties, *Advances and Applications in Electroceramics: Ceramic Transactions*, **226**, 77-90 (2011).

34. S.L. Pyshkin, Stimulated emission in gallium phosphide, Presented by Nobel Prize Laureate A.M. Prokhorov, *Sov. Phys. Dokl.,* **19**, 845-846 (1975)).

35. Sergei L. Pyshkin and John Ballato, Long-Term Convergence of Bulk- and Nano-Crystal Properties, Chapter 19 in "Optoelectronics – Materials and Technics", ISBN 978-953-307-276-0,InTech – Open Access Publisher, 459-475 (2011)

36. Zallen R. and Paul W. Band Structure of Gallium Phosphide from Optical Experiments at High Pressure, *Phys. Rev.* **134** A1628-A1641 (1964).

37. S.L. Pyshkin et al., Many-Quantum Absorption in Gallium Phosphide, *Opto-Electronics*, **2**, 245-249 (1970)

38. Kittel, C. Elastic constants and elastic waves, In: *Introduction to Solid State Physics*, the 4th edition, 133-156 (1971), John Wiley& Sons, Inc., ISBN 0471490210, New York

39. S.L. Pyshkin, Heterostructures (CaSrBa)F$_2$ on InP for Optoelectronics, Report to the US AFOSR/EOARD on the Contract No. SPQ-94-4098 (1995).

40. S.L. Pyshkin, V.P. Grekov, J.P. Lorenzo, S.V. Novikov, and K.S. Pyshkin, Reduced Temperature Growth and Characterization of InP/SrF$_2$/InP(100) Heterostructure, Physics and Applications of Non-Crystalline Semiconductors in Optoelectronics, *IATO ASI Series,* **36**, 468-471 (1996).

41. S.L. Pyshkin, CdF$_2$:Er/CaF$_2$/Si(111) Heterostructure for EL Displays, Report to the US AFOSR/EOARD on the Contract No. SPQ-97-4011 (1997).

# POROSIFICATION OF CaO-B$_2$O$_3$-SiO$_2$ GLASS-CERAMICS BY SELECTIVE ETCHING FOR SUPER-LOW κ LTCC

F. Yuan[1], Y. T. Shi[*1], J. E. Mu[1], Z. X. He[2], J. H. Guo[2], Y. Cao[2]
[1]Tianjin Polytechnic University, Tianjin, China
[2]Seven Stars Electronics Corporation, Beijing, China

## ABSTRACT

CaO-B$_2$O$_3$-SiO$_2$ glass ceramics made from Ferro A6M tape have been used to make low κ substrates in LTCC for microwave application. The raw glass in the tape could crystallize at 850°C to form wollastonite (CaSiO$_3$) and calciborite (CaB$_2$O$_4$). In microwaves, the glass ceramic has a dielectric constant κ about 5.9. To further lower the κ, the porosification of the glass ceramic is achieved in this work. A 19wt% sodium hydroxide aqueous solution could leach the silicon and boron from the glass and the CaB$_2$O$_4$ phase, respectively, without attacking CaSiO$_3$. The leached residue could be removed from an ultrasonic treatment in water. Below 40°C, the etching is dominated by the surface reaction with an activation energy E$_a$=111 kJ/mol. When the temperature goes higher, however, the reaction rate increases more quickly than the diffusion, and the rates of the two processes become comparable at least up to 80°C. Furthermore, the penetration depth varies as the square root of the etching time, suggesting that, as the penetration goes deeper, the Knudsen diffusion becomes a rate limiting step in the etching process. The porous layer has a volume porosity 46%.

## INTRODUCTION

CaO-B$_2$O$_3$-SiO$_2$ glass-ceramics has been used to make low κ substrates in Low Temperature Co-fired Ceramics (LTCC) for microwave applications.[1] The transmission lines and passive devices could be printed on each layer, and the interlayer connection could be completed through vias. The active electronic components are usually surface-mounted on the top of the multi-structure layer. As a substrate, the dielectric constant κ has to be low to ensure a high signal speed and decoupling between planes.[2] The dielectric loss also has to be low to reduce the insertion loss, and to improve the battery life and thermal management of the modules.

An increased porosity of the LTCC could further reduce the dielectric constant and loss. Moh et al.,[3] for example, dispersed mullite or other ceramic-based bubbles in the glass matrix. Kellerman and Morrison[4,5] proposed the use of hollow microspheres in the glass matrix, and Lee et al.[6] embedded air cavities between the strip conductor and the ground plane in the LTCC.

The porosity may also be generated by selectively etching the fired LTCC. Bittner et al.[7] reported that a phosphoric-based acid may selectively etch the anorthite phase in the fired DP 951 AX LTCC. The Dupont LTCC is made by firing a lead borosilicate glass pre-filled with alumina particles. The generated composite contains corundum, anorthite and alumina in the glass matrix. The etching selectively removes the anorthite phase and achieves a maximum penetration depth of 40 μm in 8 hours at 130°C. The porous layer allowed a copper microstrip ring to be patterned through sputtering, followed by lithography and chemical etching.

In this work, porosification of a CaO-B$_2$O$_3$-SiO$_2$ glass ceramics for LTCC is achieved. A 19wt% sodium hydroxide may selectively etch the calciborite and glass phase of the fired Ferro A6M tape, and develop a porous CaSiO$_3$ layer. The inorganic part of the A6M tape is a calcium borosilicate glass. By firing it at 850°C peak, wollastonite (CaSiO$_3$) and calciborite (CaB$_2$O$_4$) could be precipitated from the glass.[8] The glass ceramics have a dielectric constant 5.9 and loss 2×10$^{-3}$ from 1 GHz to 40 GHz. By mixing rules, the porous layer is expected to have a lower dielectric constant.

EXPERIMENTS

The A6M tapes were laminated isostatically at a pressure of 32 MPa at 70°C and co-fired at a peak temperature of 850°C for 30 min in a programmable tube furnace. The etching was performed in a 19wt% sodium hydroxide aqueous solution. The designed temperature of etching was controlled in an oil bath, and was made available in the range from 20°C to 100°C.

The etching rate $r_e$ is presented by the amount of LTCC etched per unit time and unit surface of exposure to the etching solution

$$r_e = -\frac{1}{A}\frac{dm}{d\tau} \tag{1}$$

where $m$ is the mass of the LTCC, $A$ is the area of the LTCC exposed to the etching solution, and $\tau$ is the etching time. In a selective etching, a porous layer could develop and the penetration rate $r_p$ could be given by the above equation divided by the density $\rho^{(2)}$ and fraction $f^{(2)}$ of the etched phases on the surface

$$r_p = \frac{r_e}{f^{(2)} \cdot \rho^{(2)}} = -\frac{\dfrac{dm}{dt}}{f^{(2)} \cdot \rho^{(2)} \cdot A} \tag{2}$$

The penetration depth $d_p$ was determined by a dye penetration test. In a 10wt% fuchsine alcohol solution, the red dye will penetrate into the open pores generated from the etching. The solution could reach the whole depth of the porous layer within an hour. The red color stays there even after the sample is taken out from the solution and rinsed with water. The penetration depth could be determined from the fractograph of the sample.

The crystalline phases were identified through X-ray diffraction (XRD) analysis. It was conducted in a 2500 TC diffractometer from Rigaku, Japan. The X-ray beam has a wavelength of 1.541Å corresponding to the Cu K$_\alpha$–line. The microstructures of the LTCC were observed before and after etched using a Quanta 200 Scanning Electron Microscope (SEM) from FEI in Holland. The SEM micrograghs may present the morphology, etchability and connectivity of each phase. From the percentage covered by each phase, the volume fraction of the phase could be calculated.

The LTCC was polished on the surface before etching to get a smooth and planar surface. Without that, the original surface profile will add up to the etched pattern, making the two difficult to separate. The polishing was started with a 2000$^\#$ sand paper, followed by 3000$^\#$ and 5000$^\#$ ones to get a mirror finish. Figure 1 illustrates the topographies before and after the polishing.

The element Ca, B and Si in the leached solution and residue were analyzed via a PS 1000 Inductively Coupled Plasma Atomic Emission Spectrometer from Leeman in U.S.. Before the test, the leached solution was diluted with water, and the leached residue was filtered, followed by dissolved in HCl acid and diluted with water. In the preparations, the relative percentages of these elements remain unchanged.

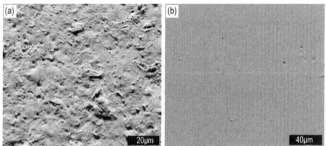

Figure 1. Surface topography of the LTCC (A6M) from SEM (a) as-fired and (b) after polished.

RESULTS AND DISCUSSION

Microstructure and phase identification

The crystalline phases of the as-fired LTCC are identified as wollastonite (CaSiO$_3$) and calciborite (CaB$_2$O$_4$) from the XRD in Figure 2(a). As reported by Ferro,[8] the glass composition of A6M tape is located at the SiO$_2$-CaSiO$_3$-CaB$_2$O$_4$ compositional triangle in the CaO-B$_2$O$_3$-SiO$_2$ phase diagram. Since the CaSiO$_3$ and CaB$_2$O$_4$ phases share a crystalline peak at $2\theta = 30°$, the highest intensity in the XRD, it is difficult to determine the major crystalline phase in the LTCC from the XRD alone. The background intensity suggests that the specimen may contain some glass phase. As the CaSiO$_3$ and CaB$_2$O$_4$ are continuously precipitated at 850°C, the composition of the residual glass approaches to SiO$_2$ by the phase diagram. The residual glass gets more acidic as the precipitation continues. Kinetically, the precipitation will slow down and eventually stop as the T$_g$ of the residual glass reaches the firing temperature.

The microstructures of the glass ceramics pre-etched for 20 hours are presented in Figure 3(a). The star-like crystals are visible in the gray, sponge-like background. The large star-like crystals have connected to form a network. The star-like crystals, by EDAX spectra, have a unity ratio of calcium to silicon, and are identified as wollastonite (CaSiO$_3$) phase. The calciborite (CaB$_2$O$_4$) is then expected to locate in the gray background. Since boron is difficult to detect by EDAX, the exact location of the CaB$_2$O$_4$ could not be surely found. Since the ratio of calcium to silicon changes along the positions tested on the gray background, the gray background may contain not only CaB$_2$O$_4$ but also some glass and nano CaSiO$_3$ crystals.

Selective etching to the LTCC

The sodium hydroxide solution is found to selectively etch the gray background. On the etched surface in Figure 3, the planarization and scratches from the polishing are still evident on the CaSiO$_3$ surface, suggesting that this phase is inert to the attacking from the etchant. The gray background, on the other hand, has been etched away, and becomes a sponge-like valley, full of voids and small light crystals in it. The thickness of the specimen remains unchanged even after 60 hours of etching at 80°C.

As presented in Figure 2, the peak intensity from the CaB$_2$O$_4$ drops after 2 hours of etching at 60°C and disappeared after 48 hours of etching at 80°C. As an X-ray usually could penetrate an inorganic oxide 0.5 to 1 μm from the surface, the peak disappearance indicates removal of the CaB$_2$O$_4$ from the etching. On the other hand, the diffraction peaks from CaSiO$_3$ remain there from the beginning to the end of the etching, indicating that CaSiO$_3$ is inert to the etchant.

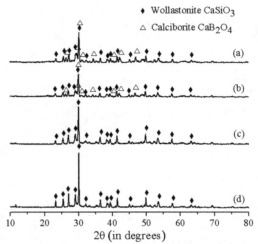

Figure 2. X-ray diffractions spectra of the LTCC (a) as-fired, (b) after etched at 60°C for 2 hours (c) after etched at 80°C for 48 hours, and (d) after etched at 80°C for 48 hours, followed by an ultrasonic treatment for 30 min.

Figure 3. SEM images of the LTCC (a) after etched at 25°C for 20 hours and (b) after etched at 25°C for 20 hours, followed by an ultrasonic treatment at 25°C for 30 min.

Even the raw glass is etchable. The tape, after binder burn-out, could dissolve in the 19wt% sodium hydroxide aqueous solution. The etching rate of the porous tape is lower than the fired ceramics. After the CaSiO$_3$ and CaB$_2$O$_4$ precipitate, the residual glass is expected to be more acidic and more etchable in the hydroxide solution.

The leached residue from the etching could be removed as "white mud" from an ultrasonic treatment. As shown in Figure 3(b), the gray background becomes open after the treatment, leaving the CaSiO$_3$ network alone there. The diffraction peaks from the CaSiO$_3$ becomes higher and narrower, suggesting that the "white mud" may contain small, unbonded CaSiO$_3$ crystals.

The silicon and boron are leached from the glass and CaB$_2$O$_4$ in the etching, leaving the calcium there as it was. As shown in table I , the leached solution is high in silicon and boron, and very low in calcium. On the other hand, the leached residue removed from the ultrasonic treatment is high in calcium and silicon. The silicon corresponds to the residual glass and the small CaSiO$_3$ crystals as a part of the "white mud" removed from the ultrasonic treatment.

Table I . Element analysis of the leached solution and residue

|  | Silicon (mol %) | Boron (mol %) | Calcium (mol %) |
| --- | --- | --- | --- |
| Leached Solution | 64.91 | 34.94 | 0.15 |
| Leached Residue | 51.53 | 2.58 | 45.89 |

The pores cover 60% of the SEM image after the ultrasonic treatment. The value is derived from a two-dimensional image. Since the LTCC is isotropic, the volume porosity is expected to be 46%, 3/2 power of the two dimensional value.

The selective etching is not conditional to the use of polishing. The sodium hydroxide solution can selectively etch the CaB$_2$O$_4$ and glass phase from the LTCC even without polishing. The polishing process was applied simply to get a smooth and planar surface so that the etched pattern could be well evident. As the porous layer will be patterned through thin film and lithography for circuit layout, the improved flatness could improve the fine line definition.[9]

Kinetics of etching

The etching process consists of two fundamental processes: surface reaction and mass transfers. The reactants transport through the diffusion boundary layer and react with the LTCC, followed by transport of the products away from the solution/LTCC interface. At equilibrium, the etching rate is given by[10]

$$r_e = -\frac{1}{A}\frac{dm}{dt} = k_e c_l = \frac{k_s k_d c_l}{(k_s + k_d)} \tag{3}$$

where $c_l$ is the reactant concentration in the bulk solution, $k_e$ is etching rate constant, $k_s$ is the reaction rate constant, and $k_d$ is the mass-transfer coefficient of the molecule in the boundary layer. Since all the three steps must occur, the slowest one, called the rate limiting step, determines the etching rate.

The porous layer was kept thin enough when the kinetics of etching is studied. At the beginning of the etching, the boundary layer may only be the etchant depletion region in the solution. As the etching goes deeper into the ceramics, the depletion region may merge to the porous layer. The etchant transport through the pores in the leached residue may be considered a Knudsen diffusion since the pore size may be much smaller than the normal mean free path. The diffusivity for a cylindrical pore is[11]

$$D_K = 97.0\,\bar{r}(\frac{T}{M_{NaOH}})^{\frac{1}{2}} \tag{4}$$

where $M_{NaOH}$ is the molecular weight of the reactive NaOH and $\bar{r}$ is the radius of the pore in the residual deposit. The diffusion flux in the solid state could be estimated by $D_K/t_p$, where $t_p$ is the thickness of the leached residue. Because of molecular collisions with the pore walls, the diffusivity is less than the normal value.

The etching rate as a function of temperature is shown in Figure 4. At a temperature lower than about 40°C, the etching rate exhibits a Boltzmann-like behavior. The slope of the linear region gives the apparent activation energy of etching $E_{app}$ about 111 kJ/mol, which is close to the enthalpy of the reaction CaB$_2$O$_4$(s) + 2OH$^-$(ad) $\rightleftharpoons$ Ca(OH)$_2$(s) + B$_2$O$_4^{2-}$(ad). Since the enthalpy is approximately equal to the activation energy of a bimolecular reaction from a transition state theory point of view,[12] the apparent activation energy corresponds to the surface reaction. In other words, $k_s \ll k_d$ and $k_e \approx k_s \sim \exp(-E_a / RT)$.

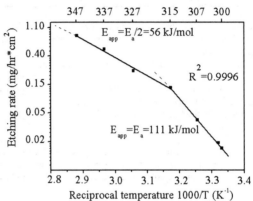

Figure 4. Arrhenius plot of the etching rate as a function of the temperature. The porous layer remains as thin as several micrometers in the test samples.

When the temperature goes higher than 40°C, the reaction rate constant $k_s$ increases more quickly and become comparable to the mass-transfer coefficient $k_d$. As shown in Figure 4, ln$k_e \sim$ -1/T still applies and the apparent activation energy $E_{app}$ = 56 kJ/mol, exactly half the activation energy of the reaction. It is a characteristic behavior of a reaction involving diffusion in small pores.

The penetration as a function of etching time is presented in Figure 5 and Figure 6. The derivative at a point on the curve gives the penetration rate at the specific moment or penetration depth. At 80°C, the penetration depth $t_p$ is linear to the time in the early stage. As the etching goes deeper than 60 μm, $t_p \sim \tau^{1/2}$ applies, and suggests that the diffusion becomes a rate-limiting step in the etching. The penetration rate $r_p$ hence drops as the etching continues. By the trend, the penetration depth will be leveling off after the etching reaches 108 μm in depth.

At 28°C, on the other hand, the etching process is dominated by surface reaction. The penetration rate is found to be 0.46 μm/hour constant at least up to 20 μm in depth. At this temperature and penetration thickness, the surface reaction is the limiting step in the etching.

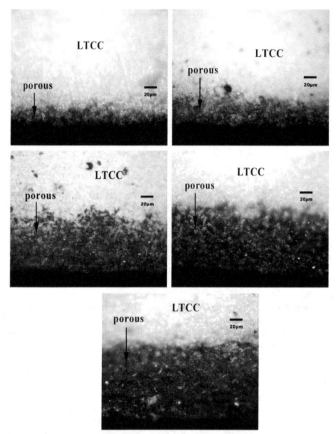

Figure 5. Fractographs of the LTCC after etched at 80°C, followed by an ultrasonic treatment and a dye penetration test both at 25°C (a)12 hours of etching, (b) 24 hours of etching, (c) 36 hours of etching, (d) 48 hours of etching, and (e) 60 hours of etching.

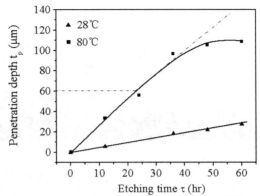

Figure 6. Penetration depth versus the etching time at 28°C and 80°C, respectively.

CONCLUSIONS

A porous layer has been developed by etching the CaO-B$_2$O$_3$-SiO$_2$ glass-ceramics in a 19 wt% sodium hydroxide aqueous solution. The etchant can selectively leach the boron from the calciborite (CaB$_2$O$_4$), and silicon from the glass in the glass ceramics. The leached residue could be removed from an ultrasonic treatment in water, leaving the wollastonite network alone there. The porous layer has a porosity about 46%.

The surface reaction has an activation energy about 111 kJ/mol. At a temperature lower than 40°C, the surface reaction controls the etching, and the etching rate exhibits a Boltzmann-like behavior. As the temperature goes up, the surface reaction rate will be comparable to the diffusion flux in the etching, where the apparent activation energy is exactly half of the reaction activation energy. Furthermore, when the porous layer develops thicker, the mass transfer will slow down, and becomes a rate limiting step in the etching, specially at elevated temperature. At 80°C, the penetration may stop at 108 μm in depth.

In the future, the etching will be performed simultaneously with the ultrasonic treatment. This may physically remove the leached residue while the etching continues. The diffusion will take place in the open pores of the CaSiO$_3$ network. The several micrometers of the pore size will speed up the mass-transfer, specially in deep penetrations, and raise the penetration limit of etching significantly. The evaluation of dielectric properties and metallization on the porous layer will also be reported.

ACKNOWLEDGMENTS
We would like to thank Ferro for supplying the A6M tapes.

*Corresponding author: yitongshi@yahoo.com

REFERENCES
[1]K. Niwa, E. Horikoshi and Y. Imanaka, Recent Progress in Multilayer Ceramic Substrates, Ceramic Transactions, *J. Am. Ceram. Soc.*, **97**, 171-182 (1999).
[2]M. I. Montrose, Printed Circuit Board Design Techniques for EMC Compliance, Institute of Electrical and Electronics Engineers, 15-45 (1996).
[3]H. K. Moh, C. D. Hoyle and C. E. Boyer, Ceramic Composite for Electronic Applications, U.S. Patent 5108958 (1992).

[4]D. Kellerman, Micro-Electronics Devices and Methods of Manufacturing Same, U.S. Patent 4865875 (1989).

[5]W. H. Jr. Morrison, Method for Preparing Ceramic Tape Compositions, U.S. Patent 4867935 (1989).

[6]Y. C. Lee and C. S. Park, Loss Minimization of LTCC Microstrip Structure with Air-Cavities Embedment in the Dielectric, *Int. J. Electron. Commun.*, **57**, 429-432 (2003).

[7]A. Bittner and U. Schmid, The Porosification of Fired LTCC Substrates by Applying Wet Chemical Etching Procedure, *J. Eur. Ceram. Soc.*, **29**, 99-104 (2009).

[8]S. K. Muralidhar et al., Low dielectric, low Temperature Fired Glass Ceramics, U.S. Patent 5164342 (1993).

[9]H. J. Levinson, Principles of Lithography, SPIE Press, 9-52 (2001).

[10]A. S. Grove, Physics and Technology of Semiconductior Devices, John Wiley and Sons, 1-366 (1967).

[11]W. L. Mccabe, J. C. Smith and P. Harriott, Unit Operations of Chemical Engineering, Fifth Edition, The McGraw-Hill, 647-685 (1993).

[12]S. R. Logan, Fundamentals of Chemical Kinetics, Longman Group, 65-84 (1996).

# MECHANOCHEMICAL BEHAVIOR OF BaNd$_2$Ti$_4$O$_{12}$ POWDER IN BALL MILLING FOR HIGH κ MICROWAVE APPLICATIONS

J. E. Mu[1], Y. T. Shi[1,*], F. Yuan[1], J. Liu[2]
[1]Tianjin Polytechnic University, Tianjin, China
[2]China Electronic Standardization Institute, Beijing, China

## ABSTRACT

BaNd$_2$Ti$_4$O$_{12}$ has a tungsten bronze type structure and a high κ in microwaves. In ball milling, a high slurry viscosity of the powder could reduce the capability of the grinding media to shear and cause impact to the particles in the milling cycles. The impact, if high, may decompose the BaNd$_2$Ti$_4$O$_{12}$. The phase stability could get improved with appropriate selections of grinding media and solvents. With 1 cm ZrO$_2$ grinding rods, a dry milling or a wet milling in xylene could cause secondary phases BaTi$_4$O$_9$, Ba$_2$Ti$_9$O$_{20}$ and Nd$_2$Ti$_4$O$_{11}$ to form. However, an acetone/IPA (50:50) mixture as a solvent may significantly drop the van der Waals interaction between particles and consequently the slurry viscosity, leading to no phase transition to the BaNd$_2$Ti$_4$O$_{12}$ powder. In addition, the phase transition may not happen if the ZrO$_2$ grinding rods are replaced with lighter grinding media, such as 1cm glass balls. The mechanical activation and phase transition also lead to high van der Waals force in the powder and thus loose structure of aggregates and poor sinterability. On the contrary, the powder pre-milled in the acetone/IPA mixture could be sintered at 1100°C or higher with the adhesive contacts of faceted grains well formed. Significant hydroxylation may take place in an aqueous milling.

## INTRODUCTION

The BaNd$_2$Ti$_4$O$_{12}$ compound has a tungsten bronze type structure, and could be used to make a high κ microwave dielectric. Upon addition and process conditions, the BaNd$_2$Ti$_4$O$_{12}$ ceramics have been reported to exhibit a dielectric constant 80 to 90, a dielectric loss 2.5×10$^{-4}$ to 4.7×10$^{-4}$ and a temperature coefficient of resonance frequency 0 to 68 ppm/°C at 3 to 4 GHz[1,2].

The mechanochemical behavior of the BaNd$_2$Ti$_4$O$_{12}$ powder is important since the powder has to get through a ball milling for grinding, dispersion or simply mixing with other ingredients in the ceramic process. Although some researchers have discovered the mechanical activation and phase transition in other kinds of materials[3-5], the mechanochemical behavior of the BaNd$_2$Ti$_4$O$_{12}$ has not been reported. By the second law of thermodynamics, a ternary compound may get activated in a milling process to become glassy form and even decompose to binary compounds until an equilibrium is reached in the between. As presented in Table I, the binary compounds are likely to have lower density and dielectric constant than the ternary compounds in BaO-Nd$_2$O$_3$-TiO$_2$ system[6-15].

In this work, ball milling is found to cause mechanical activation and phase transition to the BaNd$_2$Ti$_4$O$_{12}$ powder in certain conditions, such as the choices of the solvents and grinding media. The pre-milled particles may aggregate and perform differently in green particle packing and sintering.

Table I. Dielectric properties of compounds in BaO-Nd$_2$O$_3$-TiO$_2$ system

| Phases | κ | tanδ / 10$^{-4}$ | τ$_f$ / ppm/K | f / GHz |
|---|---|---|---|---|
| BaNd$_2$Ti$_4$O$_{12}$ | 82 to 85 | 2.5 to 4.7 | 0 to 68 | 2.89 to 3.78 |
| BaNd$_2$Ti$_5$O$_{14}$ | ~77.6 | ~8.4 | ~40 | ~3.88 |
| Ba$_2$Ti$_9$O$_{20}$ | 38 to 52 | 1.3 to 5.0 | -24 to 2 | ~4 |
| BaTi$_4$O$_9$ | 33 to 36 | 1.4 to 21 | 7 to 16 | 6 to 10 |
| Nd$_2$Ti$_4$O$_{11}$ | ~15 | ~16 | --- | --- |

## EXPERIMENTS

### Preparation of BaNd$_2$Ti$_4$O$_{12}$ powder

The single phase BaNd$_2$Ti$_4$O$_{12}$ powder was prepared via the KCl molten salt method, after Katayama et al[16]. Reagent grade BaCO$_3$, Nd$_2$O$_3$ and TiO$_2$ as starting materials were loaded with KCl in a 1.7 liter alumina jar and ball milled in isopropanol for 24 hours. The initial Ba : Nd : Ti molar ratio is 1 : 2 : 4, and the KCl was loaded 1.09 times the total starting materials. The slurry was dried at room temperature for 12 hours, and then the dried powder was heated up to 1000°C in a covered alumina crucible and held for 15 hours. The products were washed with water for 6 times so that the chloride became undetectable in silver nitrate titration.

### Ball milling of the BaNd$_2$Ti$_4$O$_{12}$

The prepared BaNd$_2$Ti$_4$O$_{12}$ powders were ball milled in a 1.7 liter milling jar. The powder load was 25 gram. The mill was filled with one third grinding media, and rolled at approximately 90 rpm for 24 hours. Several solvents, including xylene, an acetone/IPA (50:50) binary solvent and water, were charged to evaluate the milling behavior.

### Ceramic preparation

Both the raw and milled powders were used to make ceramics. Each of the powders was blended with 8wt% wax melt at 60°C. The mixture, after cooled down, was forced to get through a 50 mesh sieve. The granules were pressed to a cylindrical compact in a polished stainless steel die under a pressure of 420 MPa. The green compact has a diameter about 1.2 cm and a thickness about 0.2 cm. The sizes allow the organic burn-out to complete at 500°C in 3 hours. The raw powder and the ones pre-milled in organic solvents were sintered at 1100°C, 1150°C and 1200°C peak held for 2 hours. Since these temperatures show not high enough to the powder pre-milled in water, the water milled powder was sintered at 1300°C, 1400°C and 1500°C peaks held for 2 hours.

### Characterization

The observation of the powders morphology was made by a JSM-6700F model Scanning Electron Microscope (SEM) from JEOL in Japan. Due to the dispersion effect, the resolution of the powder image could reach 0.1 μm. It allows us to identify a single needle-like crystallite, but may not be able to distinguish a small equiaxed crystallite from another in the particle aggregates. The phase identification was completed by X-ray diffractions (XRD) with a Cu K$_\alpha$ radiation. The diffractometer is a D/MAX-2500 model from Rigaku, Japan, and the powder specimens were prepared by spreading the powder of a fixed amount into a glass supporter and pressing it to a given thickness. The sample is positioned so that the primary beam could fall on the selected area. To all the powders, a 6 kW powder rating was given, and the diffraction peak position was determined by the peak centroid method.

The slurry viscosity was measured with a Cone and Plate Viscometer. The sample was placed on the plate that was then raised to a level with small clearance from the cone. The cone was rotated at any designed speed (rpm) while the torque was measured. The angle of the cone is designed to ensure the shear stress independent of the distance from the centre of the cone. The shear rate is proportional to the rotation speed and the shear stress is related to the torque. The temperature was controlled at 25°C by passing temperature controlled water through the plate.

The wettability of the BaNd$_2$Ti$_4$O$_{12}$ was determined with a sessile drop method[17]. On the photograph of the drop, the contact angle was obtained by measuring the angle made between the tangent to the profile at the point of contact with the solid surface. The sintered plate was made by firing the compact of the raw BaNd$_2$Ti$_4$O$_{12}$ powder at 1350°C for 2 hours. To ensure the reproducibility of the BaNd$_2$Ti$_4$O$_{12}$ surface, the sintered plate was fine polished with 5000$^{\#}$ SiC sand paper and heated up to 1350°C each time before a test. The data is comparable to the results obtained from the methods designed for powders, such as the capillary height method and the displacement pressure method[17]. In the two, the powder has to be packed into a column, and it is difficult to ensure the reproducibility of the packing and to prevent the solvent from entering the column.

The fractional green density V$_s$ was used to evaluate the green density of the powder compacts. By definition, V$_s$ = ρ/ρ$_T$, where ρ is the green density and ρ$_T$ is the density of powder. The density of the powder was determined by using a MAT-5000 Auto True Denser from Seishin in Japan.

Thermal gravimetric analysis (TGA) of the powder compacts was conducted in a TG/DTA 6300 Thermal Analyzer from Seiko, Japan. The test was intended to detect any mass loss from the compacts in a temperature scanning up to 1300°C. Any mass loss in the temperature range suggests removals of the water or organic solvents from the powder surface or decomposition of the related compounds. Before the TGA, the compacts had completed the organic burnout at 500°C peak for 3 hours.

The fractional sintered density was used to determine the packing density of the sintered compact. It is defined as the sintered density divided by the theoretical density of BaNd$_2$Ti$_4$O$_{12}$ single crystal, which is about 5.15 g/cm$^3$[2]. The sintered density was determined via the Archimedes technique.

RESULTS AND DISCUSSION
Solvents, powder loading and slurry viscosity
The organic solvents and water wet the BaNd$_2$Ti$_4$O$_{12}$ and could penetrate into the three dimensional mesh structure of the powder. As presented by a sessile drop in Figure 1, the contact angle on the BaNd$_2$Ti$_4$O$_{12}$ is 29.4° with xylene, 23.9° with the acetone/IPA mixture and 21.4° with water. In fact, the solvent can spread on the particles and penetrates into the powder in a few seconds even in a static condition[18]

$$\chi^2 = \frac{d_p \gamma_{LV} t\cos\theta}{4\eta} \tag{1}$$

where $\chi$ is the depth of liquid penetration, $d_p$ is the pore size, $\gamma_{LV}$ is the liquid-vapor surface energy, $\theta$ is the contact angle, t is time, and $\eta$ is the solvent viscosity. The solvent parameters used in Equation 1 are given in Table II.

Figure 1. Contact angles of the BaNd$_2$Ti$_4$O$_{12}$ ceramics with (a) xylene, (b) acetone/IPA (50:50) binary solvent, and (c) Water.

Table II. The solvent parameters used in Equation 1

|  | xylene | acetone/IPA | water |
|---|---|---|---|
| Liquid-vapor surface energy $\gamma_{LV}$(mNm$^{-1}$) | 28.74 | 22.20 | 72.14 |
| Solvent viscosity $\eta$(cP) | 0.648 | 1.133 | 0.890 |

The type of solvents determines the loading of the BaNd$_2$Ti$_4$O$_{12}$ in a mill base. As presented in Table III, the lowest solvent charge is allowed to the mill filled with the acetone/IPA mixture as solvent. More solvent charge has to be given to the mill with water and xylene, otherwise, the slurry may pin the grinding rods to the wall of the milling jar. In a specific solvent, the slurry viscosity is related to the volume fraction of the powder as follows[19]

$$\eta_{slurry} = \eta_{solvent} \times \left(1 - \frac{\varphi}{K}\right)^{-2}$$  (2)

where $\varphi$ is the volume fraction of the powder and K is a constant at a fixed shear rate. The value is equal to the packing limit of the particle as the slurry viscosity diverges. When the shear rate rises, the K increases. The K therefore reflects the balance between the applied shear stress and the interparticle van der Waals force.

Table III. Milling Behavior of the BaNd$_2$Ti$_4$O$_{12}$ in Different Solvents

| Solvent | BaNd$_2$Ti$_4$O$_{12}$/Solvent | Slurry viscosity (pa•s) |
|---|---|---|
| Air | N/A | N/A |
| Xylene | 1 : 5.3 | $10^3$ |
| Acetone/IPA | 1 : 1.9 | $10^0$ |
| Water | 1 : 2.8 | $10^1$ |

The slurry viscosity is dominated by the potential energy due to the van der Waals interaction. For two spherical particles suspended in a solvent, the potential energy can be approximately calculated using the following equation[20]

$$V_a = -A_{S/L/S}\left(a/12H\right)$$  (3)

where a is the particle size, H is the interparticle distance, and $A_{S/L/S}$ is the combined Hamaker constant for the solid in the solvent and can be estimated from the equation

$$A_{S/L/S} = A_S + A_L - 2\left(A_S A_L\right)^{1/2}$$  (4)

where $A_S$ and $A_L$ are the Hamaker constants of the solid and solvents in vacuum. For the solid, the Hamaker constant can be calculated from the static dielectric constant $K_s(0)$ of the material

$$A_s(kT) = 113.7 \frac{\left(K_s(0) - 1\right)^2}{\left(K_s(0) + 1\right)^{3/2}\left(K_s(0) + 2\right)^{1/2}}$$  (5)

Assuming $K_s(0) = 100$ for BaNd$_2$Ti$_4$O$_{12}$, we may estimate $A_S$ to be 109 kT from Equation 5. Similar calculations have been performed for the different solvents, including air in dry milling. Therefore, the $A_{S/L/S}$ values are calculated for these slurries, and presented in Figure 2.

The $A_{S/L/S}$ values are correlated to the slurry viscosity. The $A_{S/L/S}$ value is maximum in dry milling, followed by xylene, the acetone/IPA and water. According to Equation 3, a higher $A_{S/L/S}$ requires more solvent charge to increase the interparticle distances so that the van der Waals interaction and slurry viscosity could drop.

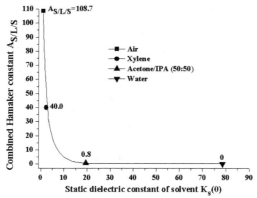

Figure 2. Combined Hamaker constant for BaNd$_2$Ti$_4$O$_{12}$ in the solvents with different static dielectric constant.

Slurry viscosity determines the type of motion in a ball mill. In the acetone/IPA binary solvent, the slurry viscosity is thin enough for the grinding media to cascade. The high slurry viscosity in dry milling or in xylene, however, reduced the capability of the grinding media to shear. The motion change could be observed, as illustrated in Figure 3, when the alumina jar was replaced with a transparent plastic jar. The media motion was found to switch from cascading to falling or cataracting, giving impact to the powders.

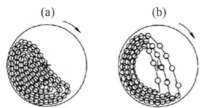

Figure 3. Types of motion in a ball mill (a) cascading, (b) falling or cataracting[21].

Significant hydroxylation may take place in the aqueous milling. The BaNd$_2$Ti$_4$O$_{12}$ gained weight after milled in water. According to the TGA data presented in Figure 4, the powder compact starts to lose weight at 600°C until 1000°C with a total loss about 55%. The temperature has significantly surpassed the thermal energy necessary to remove the bound water on the BaNd$_2$Ti$_4$O$_{12}$, and corresponds to the decomposition of hydroxides[22]. Note that the powder compact had completed the organic burnout before the TGA started. The 55% weight loss indicates that the hydroxylation is so significant that it may greatly increase volume fraction of the powder and drop the solvent volume in the mill. Both contribute to the slurry viscosity by equation 2.

Grinding, activation and phase transition
The raw BaNd$_2$Ti$_4$O$_{12}$ powder, as illustrated in Figure 5(a), is a mixture of needle-like particles and equiaxed particles. The phases, as presented in Figure 6(a), are identified as orthorhombic BaNd$_2$Ti$_4$O$_{12}$ and BaNd$_2$Ti$_5$O$_{14}$. The highest diffraction peaks correspond to (720) plane in

BaNd$_2$Ti$_4$O$_{12}$ and (270) plane in BaNd$_2$Ti$_5$O$_{14}$. The overlap of the two leads to peak broadening, making it difficult to conduct a quantitative phase analysis and to determine the crystallite size from the peak width using the Scherrer formula. As shown in Figure 5, a wet milling could grind all the needle-like particles into equiaxed ones, but in a dry milling, only part of the needle-like particles could be ground.

The equiaxed particles may aggregate and the aggregate size is solvent dependent. As shown in Figure 5, milling in xylene and water leads to more open structure than in the acetone/IPA binary solvent. As the aggregate sizes may follow a certain distribution, the average should be used as a measure of the level of the aggregation.

High level of the particle aggregation is a sign of stronger van der Waals interaction. If particles, once they have joined the aggregate, are able to rearrange, the resulting aggregate is likely to be rather compact. However, if once stuck the attraction energy is too high for the particle to move again, the aggregate will be much more open in structure, since the particles arriving at the aggregate later will tend to find access to the interior of the aggregate blocked by the earlier arrivals.

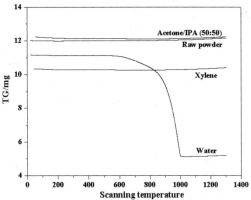

Figure 4. Thermal gravimetric analysis of the powder compacts (a) the raw BaNd$_2$Ti$_4$O$_{12}$ powder, (b) pre-milled in xylene, (c) in the acetone/IPA (50:50) binary solvent, and (d) in water. Note that the organic burnout was completed at 500 ℃ in advance.

The milling also causes activation and phase transition to the BaNd$_2$Ti$_4$O$_{12}$ powder. As shown in Figure 6, secondary phases BaTi$_4$O$_9$, Ba$_2$Ti$_9$O$_{20}$ and Nd$_2$Ti$_4$O$_{11}$ become detectable after a dry or wet milling in xylene or water. The formation of the new phases is combined with a drop in the peak intensity of the main phase and an increase of the background intensity. At $2\theta = 31.6°$, for example, the background is found to be 55 counts after milled in the acetone/IPA binary solvent, followed by 445 counts in xylene, 432 counts in water and 553 counts in dry. Since the glassy phase and the secondary binary phases both have lower density than the BaNd$_2$Ti$_4$O$_{12}$ crystalline phase, the powder density drops after ball milling. As presented in Tables IV and V, the degree of the dropping is approximately proportional to the amount of the secondary phases formed.

The mechanochemical behavior comes from the impact applied to the powder in the mill. Let the charge of the acetone/IPA binary solvent drops from 60 ml to 40 ml. It raises the slurry viscosity significantly and switches the media motion from shear to impact. By the XRD in Figure 7, the secondary phases are determined to be BaTi$_4$O$_9$, Ba$_2$Ti$_9$O$_{20}$ and Nd$_2$Ti$_4$O$_{11}$. The milling is still able to grind the needle-like particles to equiaxed ones according to the SEM micrograghs in Figure 8. Compared to the 60 ml solvent charge, the aggregates from the 40 ml solvent charge are more open, suggesting stronger interparticle attraction in the powder.

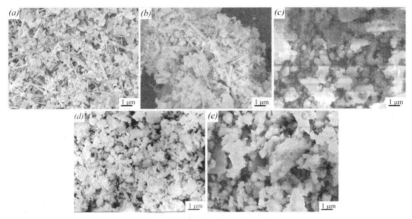

Figure 5. SEM micrographs of the BaNd$_2$Ti$_4$O$_{12}$ powders (a) the raw, (b) after dry milled for 24 hours with 1 cm ZrO$_2$ grinding rods, (c) pre-milled in xylene for 24 hours with 1 cm ZrO$_2$ grinding rods, (d) in the acetone/IPA (50:50) binary solvent, and (e) in water.

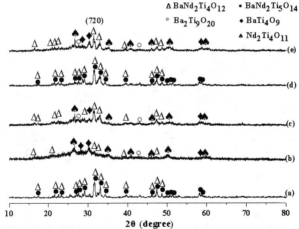

Figure 6. X-ray diffraction spectra of the BaNd$_2$Ti$_4$O$_{12}$ powders (a) the raw, (b) after dry milled with 1 cm ZrO$_2$ grinding rods, (c) after milled with 1 cm ZrO$_2$ grinding rods in xylene, (d) in the acetone/IPA (50:50) binary solvent, and (e) in water.

Table IV. Theoretical Densities of $BaNd_2Ti_4O_{12}$ and the Secondary binary Phases

| $BaNd_2Ti_4O_{12}$ | $BaNd_2Ti_5O_{14}$ | $BaTi_4O_9$ | $Ba_2Ti_9O_{20}$ | $Nd_2Ti_4O_{11}$ |
|---|---|---|---|---|
| 5.15 | 5.60 | 4.80 | 4.76 | 5.13 |

Table V. Powder Densities of the $BaNd_2Ti_4O_{12}$ Powder before and after Milled (g/cm³)

| $BaNd_2Ti_4O_{12}$ powder | Pre-milled in | | |
|---|---|---|---|
| | Xylene | Acetone/IPA | Water |
| 4.78 | 3.16 | 3.63 | 3.28 |

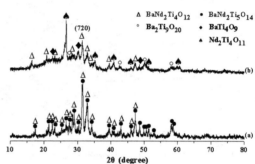

Figure 7. X-ray diffraction spectra of the $BaNd_2Ti_4O_{12}$ powders after milled with two different loads of the acetone/IPA binary solvent (a) 60ml, (b) 40ml.

Figure 8. SEM micrographs of the $BaNd_2Ti_4O_{12}$ powders after milled with two different loads of the acetone/IPA binary solvent (a) 60ml, (b) 40ml.

The intensity of the mechanical impact is determinative to cause the activation and phase transition. By replacing the 1cm $ZrO_2$ rod with 1cm glass ball, the intensity gets reduced. According to the XRD in Figure 9, the reduced intensity generates no secondary phase in the aqueous milling. By the SEM micrographs in Figure 10, the milling is still able to grind the needle-like particles into equiaxed ones. The aggregates become relatively compact, suggesting weaker van der Waals interaction in the powder. In the two cases, the shear stress approximately remains the same since the media charge and rotational rate of the two mills are the same.

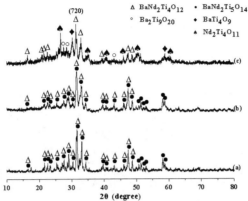

Figure 9. X-ray diffraction spectra of the BaNd$_2$Ti$_4$O$_{12}$ powders (a) the raw powder, (b) after milled in water for 24 hours with 1 cm glass grinding balls, (c) with 1 cm ZrO$_2$ grinding rods.

Figure 10. SEM micrographs of the BaNd$_2$Ti$_4$O$_{12}$ powders (a) after milled in water for 24 hours with 1cm glass grinding balls, (b) with 1cm ZrO$_2$ grinding rods.

Particle packing and sintering behavior

Table VI presents the green and sintered densities of the pressed compacts, both in fractional form. Since the limited sample number of the green compacts, it is hard to determine if the green densities are significantly different from each other or not. However, the sintered density of the raw powder is significantly higher than the pre-milled powders after sintered at 1150°C or higher. The sintered density data is consistent with the shrinkage data given in Table VII. The shrinkage of the raw powder is much higher than the pre-milled powders at 1150°C or higher. As will been seen, the high sintered density with the powder pre-milled in water is due to the porous sintered structure with the pores open and connected to each other.

The microstructures of the powder compacts sintered at several peak temperatures are presented in Figure 11. In the raw BaNd$_2$Ti$_4$O$_{12}$ powder, column-like grains are developed. With the increase of the peak temperature, the column-like grains seem to be more and more aligned along the plane of the cylindrical compact. The Ostwald ripening becomes evident at 1200°C.

Table VI. Fractional Green and Sintered Densities of Pressed Compacts

| BaNd$_2$Ti$_4$O$_{12}$ | Green density (%) | Sintered density (%) |
|---|---|---|
| The raw powder | 48.1 | 87.4 |
| Pre-milled in | | |
| Xylene | 52.8 | 63.0 |
| Acetone/IPA | 49.9 | 63.4 |
| Water | 61.6 | 88.7 |

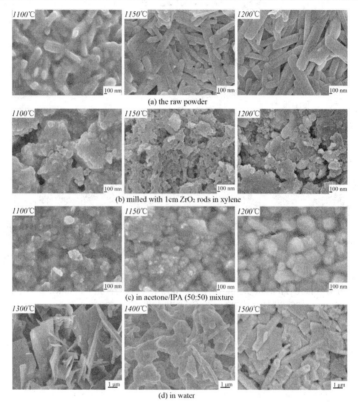

Figure 11. Microstructures of the BaNd$_2$Ti$_4$O$_{12}$ powder compacts sintered at different peak temperatures for 2 hours (a) the raw powder, (b) pre-milled with 1cm ZrO$_2$ rods in xylene, (c) in the acetone/IPA (50:50) binary solvent, and (d) in water.

In the center graphs of Figure 11 are the microstructures developed from the powders pre-milled in the acetone/IPA mixture and xylene. In both cases, the microstructures consist of fine and equiaxed grains. There is no Ostwald ripening taking place at least up to 1200°C peak for two hours. The difference between the two is that in the first case adhesive contacts of faceted grains are well formed, while in the latter breakaway from the pores in the powder compact takes place and leads to inhibited densification.

Table VII. Thickness Shrinkage of the Pressed Compacts at Different Peak Temperatures for 2 hours

| Peak Temperature(°C) | The raw powder (%) | Pre-milled in | |
|---|---|---|---|
| | | Xylene (%) | Acetone/IPA (%) |
| 1100 | 15.6 | 10.6 | 16.9 |
| 1150 | 20.7 | 15.6 | 16.0 |
| 1200 | 20.7 | 15.6 | 16.0 |

The breakaway results from the large pores in the compacts. With high level of the particle aggregation, the aggregates as clusters of small particles sinter together into dense regions, leaving difficult-to-sinter large pores between the clusters as evident in Figure 11. While the intra-aggregate pores shrink, the inter-aggregate pores grow and finally breakaway in sintering.

The low density phases also hinter the densification. In sintering, the low density phases will change back to the BaNd$_2$Ti$_4$O$_{12}$ phase with higher density, reducing the fractional density of the compacts. The phase transition usually takes place on the particle surfaces.

The powder pre-milled in water is difficult to sinter. As illustrated in the bottom graph, the microstructure is composed of randomly oriented plates with tremendous pores in it. The compacts sintered up to 1500°C peak are still porous in structure and fail in the dye penetration test, indicating that the pores are open and connected to each other.

CONCLUSIONS

Ball milling may cause mechanical activation and phase transition to the BaNd$_2$Ti$_4$O$_{12}$ powder. The secondary phases BaTi$_4$O$_9$, Ba$_2$Ti$_9$O$_{20}$ and Nd$_2$Ti$_4$O$_{11}$ may become detectable in XRD when the impact is strong. The impact energy may get reduced if the 1cm ZrO$_2$ grinding rods are replaced with 1cm glass balls. By doing that, the secondary phases are undetectable.

The grinding media is able to shear if the slurry viscosity is low enough. At a constant solvent loading, the lowest viscosity of the slurry could be achieved using an acetone/IPA (50:50) binary solvent. As the polarity of solvent matches BaNd$_2$Ti$_4$O$_{12}$'s, the van der Waals interaction for two spherical particles could reach the minimum, significantly dropping the slurry viscosity. No secondary phases are detected if the ratio of the BaNd$_2$Ti$_4$O$_{12}$ powder to the acetone/IPA binary solvent is less than 1:1.9. Higher ratio, use of xylene as a solvent or dry milling all cause the secondary phases to become detectable. Significant hydroxylation could take place in an aqueous milling, especially if 1cm ZrO$_2$ grinding rods are charged.

Since the mechanical activation causes the formation of low density glassy and secondary crystalline phases, the powder density drops. The milling also results in stronger van der Waals force and yields more open structure of the aggregates.

The powders pre-milled in the organic solvents could be sintered at 1100°C or higher, with a microstructure full of fine and equiaxed grains. The degree of mechanical activation has to be appropriate for the powder to sinter. Both the low density phases and particle aggregation hinder the sintering. Large inter-aggregate pores in the compact may break away in sintering and prohibit the densification. Without much mechanical activation, the powder pre-milled in the acetone/IPA binary solvent, however, could have the adhesive contacts of faceted grains well formed in sintering. The powder pre-milled in water is difficult to sinter at least up to 1500°C peak. The microstructure consists of randomly oriented plates with tremendous pores in it.

References

[1] E. A. Nenasheva and N. F. Kartenko, High Dielectric Constant Microwave Ceramics, *J. Eur. Ceram. Soc.*, **21**, 2697-2701 (2001).3

[2] R. Ratheesh, H. Sreemoolanadhan, M. T. Sebastian, and P. Mohanan, Preparation, Characterisation and Dielectric Properties of Ceramics in the BaO-Nd$_2$O$_3$-TiO$_2$ System, *Ferroelectrics*, **211**, 1-8 (1998).

[3] P. Balaz, I. Ebert, Thermal Decomposition of Mechanically Activated Sphalerite, *Thermochim Acta*, **180**, 117-123 (1991).

[4] K. Tkacova, P. Balaz, Reactivity of Mechanically Activated Chalcopyrite, *Int. J. Min. Proc.*, **44/45**, 197-208 (1996).

[5] V. V. Boldyrev, Mechanochemistry and Mechanical Activation of Solids, *Uspekhi Khimii*, **75**, 203-216 (2006).

[6] R. Ratheeesh, H. Sreemoolanadhan, M. T. Sebastian and P. mohanan, Preparation, Characterisation and Dielectric Properties of Ceramics in the BaO-Nd$_2$O$_3$-TiO$_2$ System, *Ferroelectrics*, **211**, 1-8 (1998).

[7] Tae-Hong Kim, Jung-Rae Park, Suk-Jin Lee, Hee-Kyung Sung, Sang-Seok Lee and Tae-Goo Choy, Effects of Nd$_2$O$_3$ and TiO$_2$ Addition on the Microstructures and Microwave Dielectric Properties of BaO-Nd$_2$O$_3$-TiO$_2$ System, *ETRI Journal*, **18**, 15-27 (1996).

[8] K. Stanly Jacob, R. Satheesh, R. Ratheesh, Preparation and Microwave Characterization of BaNd$_{2-x}$Sm$_x$Ti$_4$O$_{12}$ ($0 \leqq x \leqq 2$) Ceramics and Their Effect on the Temperature Coefficient of Dielectric Constant in Polytetrafluoroethylene Composites, *Materials Research Bulletin*, **44**, 2022–2026 (2009).

[9] T. Negas, G. Yeager, S. Bell, R. Armen, in: P. K. Davies, R. S. Roth, Jackson (Eds.), Proceedings of the International Conference on the Chemistry of Electronic Ceramic Materials, Wyoming, **21**, 17-22 (1990).

[10] H. M. O'Bryan, JR, J. Thomson and J. K. Plourde, Effects of Chemical Treatment on Loss Quality of Microwave Dielectric Ceramics, *Ber. Dt. Keram. Ges.*, **55**, 348-351 (1978).

[11] J. K. Ploude, D. F. Linn, et al., Ba$_2$Ti$_9$O$_{20}$ as A Microwave Dielectric Resonator, *J. Am. Ceram. Soc.*, **58**, 418-420 (1975).

[12] H. M. O'BRYAN, JR., and J. THOMSON, JR, A New BaO-TiO$_2$ Compound with Temperature Stable High Permittivity and Low Microwave Loss, *J. Am. Ceram. Soc.*, **57**, 450-453 (1974).

[13] D.W. Kim, D.G. Lee, K.S. Hong, Low-Temperature Firing and Microwave Dielectric Properties of BaTi$_4$O$_9$ with Zn-B-O Glass System, *Mater. Res. Bull.*, **36**, 585-595 (2001).

[14] J.H. Choy, Y.S. Han, Microwave Characteristics of BaO-TiO$_2$ Ceramics Prepared Via a Citrate Route, *J. Am. Ceram. Soc.*, **78**, 1167-1172 (1995).

[15] M. H. Weng, T. J. Liang, C. L. Huang, Lowering of Sintering Temperature and Microwave Dielectric Properties of BaTi$_4$O$_9$ Ceramics Prepared by the Polymeric Precursor Method, *J. Eur. Ceram. Soc.*, **22**, 1693-1698 (2002).

[16] K. Katayama and Y. Azuma, Molten Salt Synthesis of Single-Phase BaNd$_2$Ti$_4$O$_{12}$ Powder, *J. Mater. Sci.*, **34**, 301-305 (1999).

[17] S. Wu, Polymer Interface and Adhesion, Marcel Dekker, 257-274 (1982).

[18] A. F. Lisovsky, Thermodynamics of Isolated Pores Filling with Liquid in Sintered Composite Materials, *Mat. Trans.*, **25A**, 733-740 (1994).

[19] R. A. L. Jones, Soft Condensed Matter, Oxford University Press, 49-71 (2002).

[20] P. Somasundaran, B. Merkovic, S. Krishnakumar, X. Yu, Colloid Systems and Interfaces Stability of Dispersions through Polymer and Surfactant Adsorption, Handbook of Surface and Colloid Chemistry, Editor K. S. Birdi, CRC Press, 559-602 (1997).

[21] P. Balaz, Mechanochemistry in Nanoscience and Minerals Engineering, Springer, 118-119 (2008).

[22] D. R. Lide, ed., CRC Handbook of Chemistry and Physics, Internet Version 2007 (87th Edition), Taylor and Francis, Boca Raton, FL, 4-50 – 4-96 (2007).

EVALUATION OF ELECTROACTIVE POLYMER (EAP) CONCEPT TO ENHANCE
RESPIRATOR FACIAL SEAL

Mark Stasik, Jay Sayre, Rachel Thurston, Wes Childers, Aaron Richardson, Megan Moore
Battelle Memorial Institute
Columbus, Ohio, U.S.A.

Paul Gardner
Edgewood Chemical Biological Center
Aberdeen Proving Ground, Maryland, U.S.A.

ABSTRACT
        Advanced respirator seal concepts are being explored as part of an on-going research effort
sponsored by the Joint Science and Technology Office, Defense Threat Reduction Agency, to provide
enhanced technologies for the next generation of joint service respirators. Ionic Polymer-Metal
Composites (IPMCs) are a class of Electroactive Polymers (EAPs) that may provide the potential for
an active sealing system to enhance respirator fit. IPMCs are composite materials that bend and exert
force in response to an applied voltage of 1 to 4 volts. The objective of this work was to characterize
the force exerted by selected IPMCs and to assess the feasibility of using them in a respirator to exert a
force on the seal/skin interface to reduce leakage.

INTRODUCTION

Respirator Face Mask Sealing
        Respirators are designed to protect a wearer from breathing in various airborne toxic materials.
The protection provided can often be limited by the quality of the seal between the respirator and the
wearer's face. There is potential for leakage at this seal which may result in the wearer being exposed
to toxic fumes, because the pressure within a face mask can be below atmospheric pressure. EAPs
have the potential to help improve the sealing of face masks. The basic concept is that an EAP may be
able to exert force on the sealing flange to reduce the potential for leakage. Figure 1 illustrates a
specific example of a configuration in which an EAP could be used in a mask to exert force on a
sealing surface. The advantage of an EAP over a simple spring in this application is that several
independently operating EAPs could be installed in different locations in the mask, which can be
activated and deactivated as needed. This would allow for additional sealing forces when and where
they are needed.
        The goal of this work was to design a preliminary sealing device, similar to what is shown in
Figure 1. The first step was to determine the ability of ionic polymer-metal composite (IPMC) EAPs
to exert force in a blocked load orientation. This included measuring the force generated by the change
in shape of IPMC samples as a function of voltage, sample type, and geometry. The data from the
force tests were used to design a simple device that could be applied to a simplified leak test of a low-
vacuum system.

Background on Electroactive Polymers (EAPs)
        Electroactive polymers (EAP) are materials that change shape in response to an applied
voltage. EAPs are a broad range of materials that usually are separated into two groups. One group is
known as ionic / wet / or electrochemical EAPs. These materials change shape as a result of ionic
mass transport in response to an applied voltage. The other group is known as electronic/ dry / or
field-activated EAPs. These materials change shape in response to an electric field that is created by
an applied voltage.

147

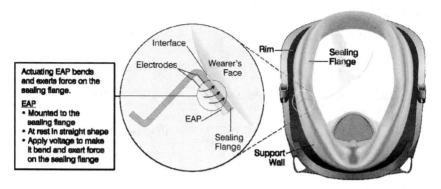

Figure 1. Example of how an EAP could be configured in a mask to exert force on a sealing surface.

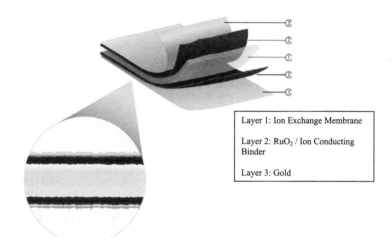

Layer 1: Ion Exchange Membrane

Layer 2: RuO₂ / Ion Conducting Binder

Layer 3: Gold

Figure 2. Layered structure of Ionic Polymer-Metal Composite (IPMC) EAP.

The specific type of EAPs investigated in this work are called ionic polymer metal composites (IPMCs). Figure 2 illustrates the basic design of IPMCs, which are part of the ionic EAP group. IPMCs consist of a center ion exchange membrane, such as Nafion®, sandwiched between two electrodes made up of $RuO_2$ and an ion conducting binder. In this work, both the ion exchange membrane and ion conducting binder in the electrodes were Nafion®. Lastly, an outer layer of gold is applied to the outer surface of each electrode to allow for uniform potential to be applied along the entire electrode area.

IMPCs bend in response to an applied voltage, as shown in Figure 3. Figure 4, which is taken from Reference 1, explains the two mechanisms that cause this bending. Mechanism 1 involves mobile cations in the ion exchange membrane migrating toward the negatively charged electrode. If the membrane has absorbed a solvent, such as water, which is illustrated in Figure 4, the solvent molecules are pulled with the mobile cations. This mass transport leads to an accumulation of material at the negative electrode, which causes expansion at that side of the membrane, resulting in bending. Ionic liquids are another type of solvent that are used in IPMCs. Ionic liquids can withstand higher voltages than water, and do not evaporate or seep out of the ion exchange membrane.

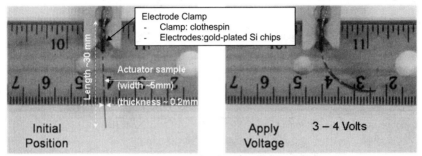

Figure 3. Example of an EAP bending.

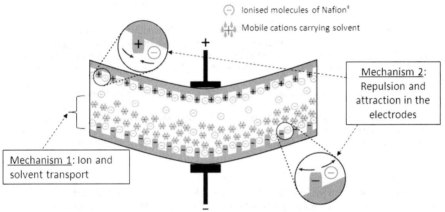

Figure 4. Two mechanisms that lead to bending in IPMCs. (Reference 1)

Mechanism 2 in Figure 4 involves repulsive and attractive forces within the electrodes. At the negative electrode, the negative charge leads to a repulsive force to the immobile anions in the ionic exchange material. This leads to an expansion effect, which contributes to the bending of the EAP. At the positive electrode, the positive charge leads to an attractive force to the immobile anions in the ionic exchange material. This leads to a contraction effect, which also contributes to the bending.

EXPERIMENTAL APPROACH

EAP Designs
    Three different IPMC EAP designs were investigated, as shown in Table I. These IPMC designs are based on previous work published by other researchers (References 2, 3,and 4).

Table I. IPMC EAP designs

| Parameter | Variation 1 | Variation 2 | Variation 3 |
|---|---|---|---|
| Center ion exchange membrane | Nafion® 115 (0.005in. = 127μm thick) | | Nafion® 117 (0.007in. = 179μm thick) |
| Electrode metal | $RuO_2$ | | |
| Electrode ion conducting binder | Nafion® | | |
| Membrane solvent | Ionic Liquid (1-Ethyl-3-methylimidazolium trifluoromethanesulfonate; CAS: 145022-44-2) | None (Dry) | |
| Outer layer gold coating | Vacuum evaporation gold coating: ~100nm gold coating: <10 Ω/cm | | |

    In design variations 1 and 2, Nafion® 115 (0.005in. = 127μm thick) is sandwiched between Nafion® binder / $RuO_2$ electrodes. Other researchers (Reference 3 and 4) generally suggest that the ionic liquid solvent (which adds fabrication steps and expense) leads to improved no-load deformation of IPMCs, but there appears to be little information on how the ionic liquid solvent affects the load-exerting capability of IPMCs. IPMCs with an ionic liquid solvent (Variation 1) and without ionic liquid (Variations 2 and 3) were made to investigate the how solvents affect the load-exerting capability of IPMCs.
    Another variation to the design (Variation 3) that may contribute to reaching higher forces is to use thicker Nafion® for the center ion exchange layer. Generally, a thicker center layer is expected to produce higher forces, but slower response. Nafion® 117 (0.007in. = 178μm thick) is a commercially available product.

Force Tests
    The force that is exerted by IPMC EAPs when voltage is applied was measured. Figure 5 illustrates the different EAP sample geometries and configurations that were investigated in the force tests. In these configurations, one end of the EAP sample was held in place by an electrode clamp (here a wooden clothespin). Within the "jaws" of the clamp are two gold-coated silicon chips that contact the opposite faces of the EAP membrane sample. A constant DC potential was applied for 60 seconds between the opposing gold surfaces of the electrode clamp, which caused the EAP to actuate. The free end of the EAP rested against a load cell sensor to measure the force generated when the EAP was actuated. (The load cell sensor is the cylinder attached to the grey metal box.) The ~0.2mm thickness of the EAP is visible as a dark line that extends from the electrode clamp. The wooden piece attached to the load cell sensor in Figures 5b and 5d provides a surface large enough for the entire 20mm EAP free length to be in contact with the load cell sensor. This wooden piece is not needed for the 5mm straight configuration (Figure 5a), because the diameter of the sensor is 5mm, allowing for full contact directly with the sensor.

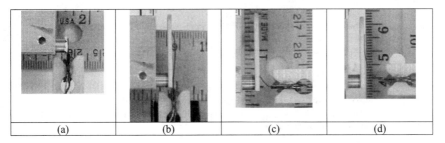

| (a) | (b) | (c) | (d) |

Figure 5. Examples of the preliminary load cell test configurations. a) Straight EAP with a short (5mm) free length; b) Straight EAP with a long (20mm) free length, c) V-bend EAP with a short (5mm) free length; d) V-bend EAP with a long (20mm) free length.

Table II. Matrix of force test conditions for each EAP design.

| Priority | Voltage | Width | Free Length | Test Configuration | # of Tests |
|----------|---------|-------|-------------|--------------------|------------|
| 1 | 3 V | 6 mm | 5 mm | Straight | 2 |
|   | 3 V | 6 mm | 5 mm | V-Bend | 2 |
|   | 3 V | 6 mm | 20 mm | Straight | 2 |
|   | 3 V | 6 mm | 20 mm | V-Bend | 2 |
| 2 | 1 V | 6 mm | 5 mm | Straight | 2 |
|   | 1 V | 6 mm | 5 mm | V-Bend | 2 |
|   | 1 V | 6 mm | 20 mm | Straight | 2 |
|   | 1 V | 6 mm | 20 mm | V-Bend | 2 |
| 3 | 3 V | 3 mm | 5 mm | Straight | 2 |
|   | 3 V | 3 mm | 5 mm | V-Bend | 2 |
|   | 3 V | 3 mm | 20 mm | Straight | 2 |
|   | 3 V | 3 mm | 20 mm | V-Bend | 2 |
| 4 | 1 V | 3 mm | 5 mm | Straight | 2 |
|   | 1 V | 3 mm | 5 mm | V-Bend | 2 |
|   | 1 V | 3 mm | 20 mm | Straight | 2 |
|   | 1 V | 3 mm | 20 mm | V-Bend | 2 |

In each force test, data recording of the force as a function of time began with the EAP not in contact with the load cell sensor. Once data recording of force started, the EAP was first gently brought in contact with the load cell sensor to the point that the load cell detected a small (~0.1 gram) mechanical pre-load. Next, a constant DC voltage was then applied for 60 seconds to actuate the EAP, and force data recording ended ~60 seconds after the voltage was turned off. This allowed for characterization of both the ramp-up in exerted force with time, as well as the decay of force in the EAP after the applied voltage was removed.

A matrix of force test conditions is shown in Table II. This matrix was designed to investigate the dependence of the force exerting capability of IPMC EAPs on the sample geometry (width and free length), configuration (straight or V-bend), and applied voltage (3V or 1V). This matrix of tests was at least partly conducted on each one of the different EAP designs, with the Priority 1 conditions considered to be the most valuable. In some cases, the lower priority conditions were not completed based on results from higher priority tests or sample availability.

Simplified Leak Tests

The data from the force tests were used to design a simple device that could be applied to simplified leak testing of a low-vacuum system. Figure 6 is a photograph of the basic components of the leak test. The components included a ~12x12x12 inch chamber, with an attached assembly of a pump, valve and gauge to create a vacuum in the chamber.

Figure 6. Photograph of the basic components of the simplified leak test.

As shown in Figure 7, on the top of the chamber there was a 3mm diameter hole that served as a designed leak passage. A 20x20mm piece of weather strip with a ~4mm hole was positioned in line with the hole in the chamber. This weather strip served as "skin" that interfaced with a piece of face mask material, although it is not clear how representative the weather strip truly is of human skin. Figure 8 illustrates the 6x6mm piece of mask material that was positioned over the hole in the skin to act as a seal to the vacuum chamber. As shown in Figure 8, the overlap of the mask material and the weather strip "skin" is only 1 – 2.25mm long around the 4mm hole. This mask/"skin" overlap is likely much lower than the overlap that would occur with a real face mask. The mask material piece weighed ~0.07 grams, which is likely to be much smaller than the typical pre-load on a mask material as a result of a user tightening the mask straps.

The simplified leak test was run by pumping the chamber down to a vacuum of 2 inches of water, turning the pump off, and monitoring the vacuum decay with time. Two inches of water may be representative of the pressure drop across a respirator canister during inhalation. The test was run with different conditions at the mask/"skin" interface. Figure 8 illustrates the condition with 0 grams (g) applied to the interface. Figure 9 illustrates the condition where either 2g, 5g, or 50g was applied to the interface using standard brass weights. Figure 10 illustrates the condition where the designed leak passage was sealed by placing a large piece of duct tape over the passage. These controlled tests were run to serve as a reference for the results of applying a simple force-exerting device involving an EAP to the leak passage.

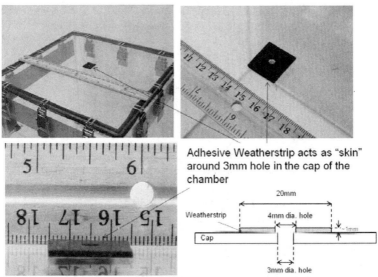

Figure 7. Designed leak passage in the top of the chamber.

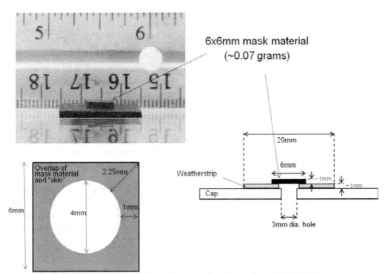

Figure 8. Interface of mask material and weather strip "skin".

Figure 9. Application of known weights to the mask material/"skin" interface.

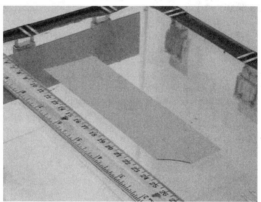

Figure 10. "Sealed" condition is simulated by placing duct tape over the weather strip.

As is explained further in the Results section, the force tests indicated that the short (5mm) free length EAP in the straight configuration (Figure 5a) generally exerted the largest force. Therefore this configuration was used as the template as a simple device, per the Figure 1 schematic, that could assist with sealing in the simplified leak test.

Figure 11 illustrates a schematic of an EAP/electrode clamp assembly that is positioned to exert a force on the mask/"skin" interface when it is actuated. Figure 12 shows a series of photographs of the EAP device applied to the leak test. The EAP/electrode clamp assembly was firmly held in place by a wooden stick that was grasped by a ring stand. Shims were used to prevent the secured EAP device from pushing down on the mask material prior to applying voltage.

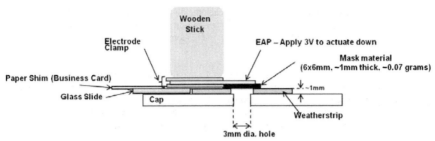

Figure 11. Schematic of EAP design that was applied to the leak test.

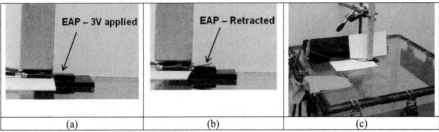

Figure 12. Photographs of the EAP positioned to apply force to the mask/"skin" interface in the simplified leak test. (a) 3V applied to the EAP; (b) EAP bent up in a retracted position; (c) View of the vacuum chamber with the EAP assembly set up on top.

RESULTS

Force Tests

Not all of the results from the force tests described in Table II for the three IPMC design variations will be presented here. Although there are many observations that could be discussed, the key observation made is that the 6mm wide, 5mm free length samples in the straight configuration at 3V generally lead to the highest force, by a wide margin, for each of the three EAP designs. Also worth noting is that the 1V tests led to less force than the 3V tests for a given geometry and configuration, as was suspected, and the 3mm wide samples exerted approximately half of the force of the 6mm wide samples.

Figures 13, 14, and 15 show the force vs. time measurements for the three IPMC designs in the straight configuration. These samples had a 6mm width and a 5mm free length, and were actuated with a potential of 3V. It was these conditions that led to the highest exerted forces for the three IPMC designs. Table III directly compares the maximum forces exerted by the three different EAPs, which can be observed in Figures 13, 14, and 15. Variation 2 (N115, dry) (Figure 14) reached the highest force in the shortest time. Therefore, this was the configuration chosen to use in the simplified leak tests. Variation 1 (N115, ionic liquid) (Figure 13) did not appear to sustain its maximum force. Variation 3 (N117, dry) (Figure 15) reached ~3.1g at 60 seconds, and was still rising.

Table III also shows a comparison of the rate that the force ramps up in the first 10 seconds of the test for these same samples in the same test conditions. Variation 1 (N115, ionic liquid) appears to have the fastest response, even though it does not reach the highest maximum force in Table III. Variation 2 (N115, dry) has roughly 75% of the force ramp-up rate of Variation 1 in the first 5 seconds. Variation 3 (N117, dry) has by far the slowest response time, which is consistent with it being the thickest membrane.

Table III. Comparison of the rate that forces ramp up for the different EAP designs in the straight, 5mm free length, 6mm wide, 3V condition. The time is with respect to the initial application of the 3V actuating potential. The pre-load in these reported forces is subtracted out.

| Variation | Force at 2 seconds | Force at 5 seconds | Force at 10 seconds | Force at 30 seconds | Force at 60 seconds |
|---|---|---|---|---|---|
| Variation 1 (N115, ionic liquid) | 0.77g | 1.50g | 1.97g | 2.43g | 2.37g [a] |
| Variation 2 (N115, dry) | 0.51g | 1.18g | 1.81g | 2.96g | 3.34g [b] |
| Variation 3 (N117, dry) | 0.16g | 0.39g | 0.86g | 2.08g | 3.10g [c] |

a) – Force decreases after a ~2.5g peak
b) – Plateau in force.
c) – Force still rising at 60 sec.

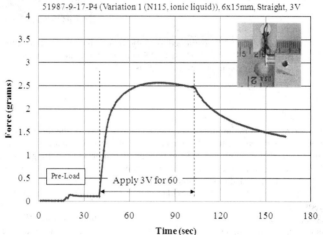

Figure 13. Force vs. time measurements for Variation 1 (N115, ionic liquid) in the straight configuration when actuated with a potential of 3V.

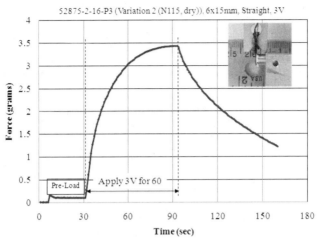

Figure 14. Force vs. time measurements for Variation 2 (N115, dry) in the straight configuration when actuated with a potential of 3V.

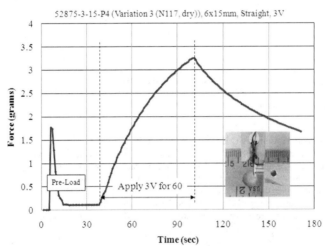

Figure 15. Force vs. time measurements for Variation 3 (N117, dry) in the straight configuration when actuated with a potential of 3V.

Simplified Leak Tests

Figure 16 shows the vacuum decay curves for the reference conditions (sealed, standard weights), as well as for the test with the EAP seal device (Figure 11 and 12). The EAP used was the Variation 2 design (N115, dry), in the 5mm free length, 6mm wide, 3V condition. The EAP curve falls between the curves for the 2g and 5g standard weight tests. This is consistent with the observation that this EAP exerts ~3g in the force test.

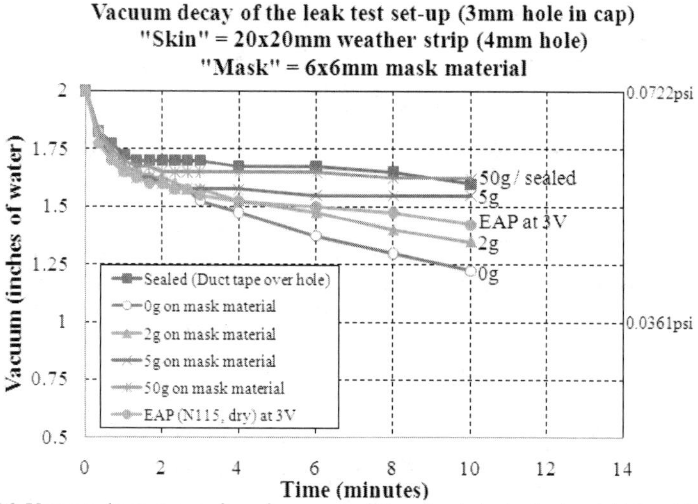
Vacuum decay of the leak test set-up (3mm hole in cap)
"Skin" = 20x20mm weather strip (4mm hole)
"Mask" = 6x6mm mask material
Vacuum (inches of water)
Time (minutes)
0.0722psi
0.0361psi
50g / sealed
5g
EAP at 3V
2g
0g
Sealed (Duct tape over hole)
0g on mask material
2g on mask material
5g on mask material
50g on mask material
EAP (N115, dry) at 3V

Figure 16. Vacuum decay curves for reference conditions (sealed, standard weights) and EAP seal device.

CONCLUSIONS

The load cell tests on IPMC EAPs revealed that, in general, the straight configuration with the short 5mm free length and 6mm wide samples tends to lead to the greatest maximum force. Of the three IPMC designs, Variation 2 (N115, dry) appears to reach the largest maximum force in the shortest time. Despite this, in the first 5-10 seconds of the test, Variation 1 (N115, ionic liquid) builds force faster than the other designs. Using the Variation 2 (N115, dry) IPMC in the leak test led to a vacuum decay between the curves for the 2g and 5g standard weights, which is consistent with the 3g force measured in the force test. This demonstrates that IPMCs have potential to help reduce leaks in low vacuum systems.

ACKNOWLEDGMENTS

We would like to thank the Joint Science and Technology Office at the Defense Threat Reduction Agency for sponsoring this work.

We would like to thank the Edgewood Chemical and Biological Center for their support and guidance.

REFERENCES

[1]C. Bonomo, L. Fortuna, P. Giannone, and S. Graziani, A Circuit to Model the Electrical Behavior of an Ionic Polymer-Metal Composite, *IEEE Transactions on Circuits and Systems – I: Regular Papers*, Vol. 53, No. 2, 338-350 (2006).

[2]M. Shahinpoor, Y. Bar-Cohen, T. Xue, J. Simpson, and J. Smith, Ionic Polymer-Metal Composites (IPMC) As Biomimetic Sensors and Actuators, *Proceedings of SPIE's 5th International Symposium on Smart Structures and Materials*, Paper No. 3324-27, 1-17 (1998).

[3]B. Akle, M. Bennett, and D. Leo, High-Strain Ionomeric – Ionic Liquid Composites via Electrode Tailoring, *Proceedings of the International Mechanical Engineering Conference and Exposition 2004*, IMECE2004-61246, 1-8 (2004).

[4]T. Mirfakhrai, J. Madden, and R. Baughman, Polymer Artificial Muscles, *Materials Today*, Vol. 10, No. 4, 30-38 (2007).

EFFECT OF SPARK PLASMA SINTERING ON THE DIELECTRIC BEHAVIOR OF BARIUM
TITANATE NANOPARTICLES

T. Sundararajan[1,*], S. Balasivanandha Prabu[2], S. Manisha Vidyavathy[1]

[1]Department of Ceramic Technology, A. C. Tech. Campus, Anna University, Chennai-600 025,
Tamilnadu, India.

[2]Department of Mechanical Engineering, College of Engineering Guindy, Anna University, Chennai-
600 025, Tamilnadu, India.

ABSTRACT
        Nano-size Barium Titanate (BT) was synthesized by the mechanochemical technique.
Synthesized particles were compacted and sintered by the conventional sintering and Spark Plasma
Sintering (SPS) techniques at 1000°C. The total sintering schedule employed was 5 minutes during
SPS and 8 hours during conventional sintering. The particles were characterized by XRD, SEM, BET
and Laser Raman Spectroscopy techniques. Dielectric studies were done using LCZ meter. Powders
sintered by SPS showed superior properties with relatively higher dielectric constant, which may be
attributed to the relatively higher tetragonality and residual stresses in the plasma sintered BT.

INTRODUCTION
        Barium Titanate (BT) is one of the traditional dielectric and ferroelectric material utilized for
the fabrication of ceramic capacitors. Over the past several decades, extensive research works have
been performed to improve the charge storage efficiency of BT based ceramic capacitors. Nano-size
BT has been eagerly desired for the miniaturization of multi layered capacitors [1, 2]. High energy ball
milling technique is considered to be one of the highly efficient processing methodologies for the
large-scale production of BT nanoparticles [3]. The grain size of the sintered BT nanoparticles dictates
its dielectric behaviour. It has been reported that the room-temperature relative dielectric constant
increases as the grain size of the BT based ceramics is decreased [4-6]. Though the nano-size BT can be
synthesized by the high energy ball milling technique, retention of nanostructure after sintering at
higher temperatures is a major research challenge. Conventional sintering of BT under ambient
conditions at high temperatures results in the exaggerated grain growth which is detrimental to its
dielectric behaviour and component miniaturization. It consumes higher amount of energy and time.
Alternative sintering methodologies are the need of the hour, in order to reduce the grain growth,
thereby enhancing the dielectric properties of BT ceramics. BT can easily be sintered to high densities
at a rapid rate in a shorter time period by the application of microwave energy due to its multiple
benefits [7, 8]. But the microwave assisted sintering seems to cause a major problem of thermal runaway
[9, 10], due to the non-uniform exposure to microwave radiation paving way to the differential
densification and selective local overheating of the compact. Spark Plasma Sintering (SPS) technique
is believed to be a practical alternative methodology for the sintering of nano-BT. It has got relatively
higher energy efficiency than the conventional and microwave sintering techniques [11, 12]. In SPS
Technique, the sample powder is pressed uniaxially in a graphite based die and punch setup. Direct
current pulse voltage is applied along the axial direction to the compact. The spark discharge occurs
around all the particles, so that the compact could be sintered uniformly and volumetrically to very
high theoretical densities in a shorter time period [13]. The problem of differential densification is
completely eliminated in the SPS process. It can be anticipated that the nanostructure of the starting
powders will be retained in the sintered compact due to the rapid heating & cooling rates and shorter
holding time employed during SPS process. In this paper, we compare the preliminary characterisation

results of the conventionally sintered and the spark plasma sintered BT nanoparticles with due emphasis in their influences on the room temperature dielectric properties of BT.

EXPERIMENTAL METHODS

Barium Hydroxide (99% purity, Loba Chemicals Ltd.) and Rutile phase Titania (99% pure, Merck Laboratories) were used as the primary precursors for the synthesis of nano-BT. These precursors of stoichiometric quantities were subjected to high energy ball milling in an indigenously developed ball mill. Tungsten Carbide (WC) balls of ~15 mm diameter were used as the milling media. Powder to ball ratio used in the milling process was 1:18 (on weight basis). Dry milling of the precursors was performed with a constant milling speed of 220 rpm for 16 hours. After milling, the powders were calcined at $750^0$C in a conventional box furnace. The calcined powders were ball milled again for 6 hours and the milled powders were divided into two equal parts. One part was sintered by SPS Technique at $1000^0$C with an overall sintering schedule of 5 minutes and the other part was sintered in a conventional box furnace at the above mentioned temperature with an overall sintering schedule of 8 hours.

CHARACTERISATIONS

The phase purity of the BT was determined by the X–ray powder diffraction (XRD) technique by using a Seifert (JSO-DEBYEFLEX 2002) Diffractometer with Cu-$K_{\alpha}$ radiation ($\lambda$=0.1540nm). The Laser-Raman spectroscopic studies were performed using the 488 nm excitation line of an argon ion laser. Surface Area measurements were carried out based on the Braunauer-Emmett-Teller (BET) theory using Nova 4000e, Surface Area & Pore size analyzer, Quantachrome Instruments. Scanning Electron Microscopy (SEM) analysis and Energy Dispersive X-Ray (EDX) analysis were carried out using Variable Pressure SEM instrument (Hitachi,Model:S-3400N) and Field Emission Scanning Electron Microscope (FE-SEM) (Zeiss Ultra 55 plus). Dielectric studies were carried out using LCZ Meter (N4L PSM 1700, HIOKI 3535)

RESULTS AND DISCUSSIONS

X-Ray Diffraction (XRD) patterns of the conventionally sintered BT and spark plasma sintered BT are shown in Figure 1.A) and Figure 1.B), respectively. The diffraction patterns of both the samples showed that the crystal structure was tetragonal in nature [JCPDS Standards (File No: 79-2265)].

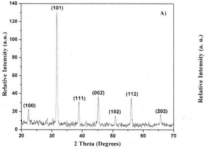 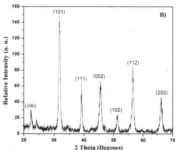

Figure 1. XRD pattern of the: A) Conventionally sintered BT, B) Spark plasma sintered BT.

It was very clear from the XRD patterns that both the samples showed a high level of crystallinity without any peaks corresponding to second/ impurity phases. The extent of tetragonality (c/a ratio) of the BT is an important parameter, which directly affects its dielectric characteristics [14, 15]. The Multi Layered Ceramic Capacitor (MLCC) manufacturing industries prefer BT powders with a tetragonality value higher than 1.008 for high capacitance applications [16]. In the conventionally sintered BT, the c/a ratio was 1.0028. In the spark plasma sintered BT, the c/a ratio was 1.032. It can be inferred that the extent of tetragonality was relatively higher in the spark plasma sintered BT. Spark plasma sintered BT showed a broader full width at half maximum (FWHM)value than that of the conventionally sintered BT, which confirms the reduced crystallite size obtained during plasma heating. The average crystallite size calculated using the Scherrer's formula [17] was found to be 40 nm for the spark plasma sintered particles and 110 nm for the conventionally sintered particles. The Laser-Raman spectra of the conventionally sintered and spark plasma sintered BT nanoparticles are shown in Figures 2.A) and Figure 2.B), respectively.

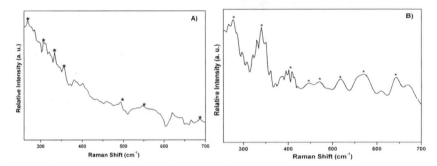

Figure 2. Laser-Raman Spectra of the: A) Conventionally sintered BT, B) Spark plasma sintered BT. (Characteristics tetragonal peaks are indexed with '*')

Raman spectra of the spark plasma sintered BT showed highly intense peaks with less noise in the spectra. Eight Raman active vibration modes were anticipated for BT having tetragonal crystal system [18]. The most predominant peaks near 516 cm⁻¹, 252 cm⁻¹, 308 cm⁻¹ and 714 cm⁻¹ were indexed with (*). The peaks around 521 cm⁻¹ and 251 cm⁻¹ were assigned to the transverse component of the optical mode of A1 symmetry and the peak at 308 cm⁻¹ was attributed to the B1 mode, indicating an asymmetry within the $TiO_6$ octahedra of BT. The broadband around 714 cm⁻¹ was related to the highest frequency longitudinal optical mode of A1 symmetry[19]. Out of the four indexed peaks, the peak around 307 cm⁻¹ is of much importance in determining the extent of tetragonality of the BT [20]. In the conventionally sintered BT, the relative intensity of the peak at 308 cm⁻¹ was relatively lesser. In the case of spark plasma sintered BT, the peak at 308 cm⁻¹ had shifted to higher wavenumbers. The reason for peak shifting could be attributed to the unrelieved stresses in the crystallites induced during the high energy ball milling process. The high intensity peak at 308 cm⁻¹ indicates the enhanced tetragonality of the spark plasma sintered BT when compared with that of the conventionally sintered BT. Thus the Raman spectral data justified the enhanced tetragonality value of spark plasma sintered BT calculated from the XRD pattern. The SEM micrographs of the conventionally sintered BT and the spark plasma sintered BT are shown in Figure 3.A) and Figure 3.B), respectively. Lineal-Intercept method (Section 10 of ASTM E 112) was used to calculate the average grain size value from the SEM micrographs. From the SEM micrographs, the average grain size was found to be 107.4 nm for the conventionally sintered BT and 44 nm for the spark plasma sintered BT.

Figure 3. SEM micrograph of the: A) Conventionally sintered BT, B) Spark plasma sintered BT.

The lower value of the average grain size of the spark plasma sintered BT was attributed to the rapid heating rate, shorter holding period and rapid cooling rate followed during the plasma heating process. The grains were of almost uniform sizes with distorted tetragonal morphology. Large numbers of coarsened regions were noticed in the SEM image of the conventionally sintered BT, which may be attributed to the longer sintering schedule followed during the conventional sintering process leading to the enhanced surface diffusion processes. BET method was used to determine the surface areas of the conventionally sintered and spark plasma sintered BT. Surface area of the spark plasma sintered BT was found to be ~17 $m^2/g$, which is relatively higher than the value obtained for conventionally sintered BT nanoparticles. Room temperature dielectric studies were carried out at 1 kHz on the conventionally sintered and spark plasma sintered BT. Samples were electroded using platinum paste and cured at $800^0C$ before carrying out the dielectric studies. Spark plasma sintered BT nanoparticles showed a relatively higher value of room temperature relative dielectric constant value ($\varepsilon_r = 3648$) than that of the conventionally sintered BT. Spark plasma sintered BT nanoparticles showed a 48% increase in the relative dielectric constant value. The dielectric loss value at the lowest frequency was lesser in the spark plasma sintered BT (tan $\delta = 0.1720$) than that of the conventionally heated BT (tan $\delta = 0.2178$). Higher relative dielectric constant value showed by the spark plasma sintered BT was attributed to its superior tetragonality and finer grain size. Internal stresses induced in the BT nanoparticles during the high energy ball milling process and the lattice strain associated with the higher cooling rates employed in the spark plasma sintering were also believed to be the reasons for superior dielectric characteristics of spark plasma sintered BT.

CONCLUSIONS

Barium Titanate (BT) nanoparticles were synthesized by the mechanochemical process. The obtained particles were sintered by conventional sintering and Spark Plasma Sintering (SPS) techniques. Spark plasma sintered BT showed superior extent of tetragonality of 1.032 as calculated from the XRD pattern. Highly intense peaks with apparent peak broadening were clearly noticed in the XRD pattern of the spark plasma sintered BT. Laser-Raman spectral data corroborated well with the results obtained from the XRD studies. Shifting of the characteristic tetragonal peak around 308 cm[-1] in the Raman spectra was attributed to the unrelieved lattice stresses induced during the high energy ball milling process. Highly uniform microstructure with distorted-tetragonal morphology of the grains was evident from the SEM micrograph of the spark plasma sintered BT. Locally coarsened grains were noticed in the SEM micrograph of the conventionally sintered BT. The surface area was relatively higher in the spark plasma sintered BT due to its finer grain size. Spark plasma sintered BT nanoparticles showed a 48% increase in the room temperature relative dielectric constant value. The dielectric loss value was relatively less in the plasma sintered BT. SPS is considered to be highly energy efficient as it had consumed 1/35th of the power consumed by the conventional sintering route. It has also enhanced the dielectric properties of BT to a very great extent.

## ACKNOWLEDGEMENT

The corresponding author[*] (T. Sundararajan, email ID: tsrajan1991@gmail.com; Ph: +91-9940350839) of the paper thanks the Centre for Technology Development and Transfer (CTDT), Anna University, Chennai, India for funding this project (File No:7672/CTDT-1/2010) under the "Students Innovative Research Projects Support Scheme" of Anna University. The support of Miss. S. Jothimani and Mr. S. Arvind Joshua Jayadev of Materials Science and Engineering Division, Department of Mechanical Engineering, College of Engineering Guindy, Anna University, Chennai is gratefully acknowledged.

## REFERENCES

[1]H. F. Kay, H. J. Vellard and P. Vousden, Atomic Positions and Optical Properties of Barium Titanate, *Nature*, **163**, 636 637 (1949).

[2]B. D. Stojanovic, Mechanochemical synthesis of ceramic powders with perovskite structure, *J. Mater. Proc. Tech.*, **143**, 78-81 (2003).

[3]I. J. Clark, T. Takeuchi, N. Ohtori and D. C. Sinclair, Hydrothermal synthesis and characterisation of BaTiO₃ fine powders: precursors, polymorphism and properties, *J. Mater. Chem.*, **9**, 83–91 (1999).

[4]H. Mostaghaci and R. J. Brook, Microstructure Development and Dielectric Properties of Fast-Fired BaTiO₃ Ceramics, *J. Mater. Sci.*, **21**, 3575-3580 (1986).

[5]L. Curecheriu, M. T. Buscaglia, V. Buscaglia, Z. Zhao and L. Mitoseriu, Grain size effect on the nonlinear dielectric properties of barium titanate ceramics, *Appl. Phys. Lett.*, **97**, 242909-242911 (2010).

[6]B. D. Stojanovic, C. R. Foschini, M. A. Zaghete, F. O. S. Veira, K. A. Peron, M. Cilense and J. A. Varela, Size effect on structure and dielectric properties of Nb-doped Barium Titanate, *J. Mater. Proc. Tech.*, **143**, 802-806 (2001).

[7]D. K. Agrawal, Microwave processing of ceramics: A review, *Current Opinion in Solid State & Mater. Sci.*, **3**, 480-486 (1998).

[8]K. H. Brosnan, G.L. Messing and D. K. Agrawal, Microwave sintering of alumina at 2.45 GHz, *J. Am. Ceram. Soc.*, **80**, 1307-1312 (2003).

[9]E. T. Thostenson and T. W. Chou, Microwave processing: fundamentals and applications, *Composites Part A*, **30**, 1055-1071 (1999).

[10]G. A. Kriegsmann, Thermal runaway in microwave heated ceramics. A one dimensional model, *J. Appl. Phys.*, **71**, 1960-1966 (1992).

[11]J. Liu, Z. Shen, M. Nygren, B. Su and T. W. Button, Spark plasma sintering behavior of nano-sized (Ba,Sr)TiO3 powders: Determination of sintering parameters yielding nanostructured ceramics, *J. Am. Ceram. Soc.*, **89**, 2689-2694 (2006).

[12]Z. Shen, M. Johnsson, Z. Zhao and M. Nygren, Spark plasma sintering of alumina, *J. Am. Ceram. Soc.*, **85**, 1921-1927 (2002).

[13]M. Omori, Sintering, consolidation, reaction and crystal growth by the spark plasma system (SPS), *Mater. Sci. Eng. A*, **287**, 183-188 (2000).

[14]D. H. Yoon and B. I. Lee, BaTiO₃ properties and powder characteristics for ceramic capacitors, *J. Ceram. Proc. Res.*, **3**, 41-47 (2002).

[15]W. Sun, C. Li, J. Li and W. Liu, Microwave-hydrothermal synthesis of tetragonal BaTiO₃ under various conditions, *Mater. Chem. Phys.*, **97**, 481-487 (2006).

[16]S. W. Kwon and D. H. Yoon, Effects of heat treatment and particle size on the tetragonality of nano-sized Barium Titanate powder, *Ceram. Int.*, **33**, 1357-1362 (2007).

[17]A. Patterson, The Scherrer Formula for X-Ray Particle Size Determination, *Phys. Rev.*, **10**, 978-982 (1939).

[18]R. Asiaie, W. Zhu, S.A. Akbar and P.K. Dutta, Characterization of submicron particles of tetragonal BaTiO₃, *Chem. Mater.*, **8**, 226 (1996).

[19]F. Maxim, P. Ferreira, P. M. Vilarinho and I. Reaney, Hydrothermal synthesis and crystal growth studies of BaTiO₃ using Ti nanotube precursors, *Cryst. Growth Des.*, **8**, 3309-3315 (2008).

[20]U. A. Joshi, S. Yoon, S. Balk and J. S. Lee, Surfactant-free hydrothermal synthesis of highly tetragonal Barium Titanate nanowires: A Structural investigation, *J. Phys. Chem. B*, **110**, 12249-12256 (2006).

# RELATIONSHIP BETWEEN ORDERING RATIO AND MICROWAVE Q FACTOR ON INDIALITE/CORDIERITE GLASS CERAMICS

Hitoshi Ohsato[1,2,4,*], Jeong-Seog Kim[1,*], Ye-Ji Lee[3], Chae-Il Cheon[3,*], Ki-Woong Chae[3*], Isao Kagomiya[4]

[1]BK21 Graduate School, Hoseo University, Asan-si Chungnam, Korea.
[2]Department of Research. Nagoya Industrial Science Research Institute, Nagoya, Japan.
[3]Department of Material Science and Engineering, Hoseo University, Asan-si Chungnam, Korea.
[4]Material Science and Engineering, Nagoya Institute of Technology, Nagoya, Japan

## ABSTRACT

Millimeterwave dielectrics for wireless communications with high data transfer and radar system for Pre-Crash Safety System are expected to be developed on the microwave Q factor. Cordierite ($Mg_2Al_4Si_5O_{18}$), is a candidate for millimeter-wave dielectrics. Cordierite has two polymorphs due to ordering of Si and Al ions on $(Si/Al)O_4$ tertahedra: disordered high symmetry form called as indialite and ordered low symmetry form cordierite. It is well known that indialite is not synthesized by solid state reactions but crystallized from glass. In this paper, the crystal structures of cordierite glass ceramics composed by the two phases are analyzed by Rietveld method, and the site occupancies are confirmed by the volumes and covalencies based on the crystal structure. It was clarified that glass ceramics with high amount of disordered crystal structure has higher Q than one with ordered one. The properties: Qf = 148,919 GHz, $_r$ = 4.7 and TCf = -29 ppm/°C. This results are consistent with our insistency that is high symmetry brings high Q more than ordering.

## INTRODUCTION

Millimeter-wave wireless communications with high data transfer rate for non-compressed digital video transmission system, and radar for Pre-Crash Safety System have been developed recently. These systems for millimeter-wave require substrates with high quality factor (Q), low dielectric constant ($_r$) and near zero temperature coefficient of resonance frequency (TCf) [1]. They also requires another physical properties such as high thermal conductivity and low thermal expansion. Millimeter-wave dielectrics are required to have high Qf value because the high frequency brings high losses. And the dielectric constants with reduced size feature should be small because of accuracy control of the fabrications. Since the substrates in radar system are exposed to a wide range in temperature inside a narrow space between front of engine room and the radiator, the TCf of substrates should be tuned to near zero. Ceramics substrates are superior more than resin substrates, because of high Q and near zero TCf, and high thermal conductivity and low thermal expansion.

We presented high Q forsterite and willemite ceramic substrates with low dielectric constant, near zero TCf, high thermal conductivity, and low thermal expansion[2]. We are also researching indialite/cordierite glass ceramics with low dielectric constant for a candidate for the substrate. The materials show high Qf values of > 200,000 GHz, low dielectric constant $_r$ = 4.7 and TCf = -27 ppm/°C, and indialite with disordering structure brings high Q more than cordierite with ordering structure[3,4]. Indialite and cordierite are polymorphs, the former is high symmetry hexagonal form with P6/mcc (No. 192), and the latter is low symmetry orthorhombic form with Cccm (No.66). We also

167

presented that the high $Q$ is brought by high symmetry instead of ordering on the materials with order-disorder transition such as Ba(Zn$_{1/3}$Ta$_{2/3}$)O$_3$ (BZT) and Ba(Zn$_{1/3}$Nb$_{2/3}$)O$_3$ (BZN) based on Koga's new theory[5-7] which is depending on the ceramic microstructure such as density, grain boundary, and lattice defect caused by non-stoichiometry. They reported some experimental evidences. For an example[7], BZT with high density obtained by spark plasma sintering for a short time shows high $Q$ same as BZT sintered by conventional solid state sintering. The BZT with high density was disorder structure. For another example[8], BZN with order-disorder transition at 1350 °C shows disordering structure with high $Q$ above the transition temperature. In accordance with the first example, even though the BZN with disorder structure transforms to the order structure by annealing at lower temperature 1200 °C, the $Q$ did not increased. So, we suggest that the high $Q$ on compounds with order-disorder transition is brought by high symmetry instead of ordering[9].

Based on our knowledge, indialite can be considered as an intermediate phase during crystallize from glass to cordierite, because indialite/cordierite ceramics were synthesized only through the crystallization process of glass which was casted in graphite mold[3]. This glass ceramic has two difficult problems to be handled during the manufacturing process, that is, deformation and cracking. The deformation due to softening above the $Tg$ could be lessened by increasing heating rate to about 600 °C/h by avoiding formation of glass phase at endothermic peak around 1200 °C in the DTA curve. The cracking depending on the anisotropic crystal growth could be avoided to a certain extent by slow cooling rate[3].

In this paper, glass ceramics of indialite/cordierite with high $Q$ are fabricated by optimizing the crystallization conditions. The amount of indialite and cordierite was obtained by using Rietveld method, and the site occupancies of tetrahedra were confirmed by the volumes and covalencies based on the crystal structure. It was clarified that glass ceramics with high amount of disordered crystal structure has higher Q than one with ordered one.

EXPERIMENTAL PROCEDURE

Mixed powders with cordierite/indialite composition were prepared by using high purity raw materials: MgO, Al$_2$O$_3$ and SiO$_2$ (> 99.9 wt% purity). The mixed powders were melted and clarified at 1550 °C, and casted in graphite mold of    10 x 40 mm. The glass rods were annealed for relieving the thermal stress in glass at 760 °C for 1 h under the glass transition temperature of 780 °C. Glass pellets with    10 x 6 mm were cut and crystallized at 1200, 1300 and 1400 °C for 10 h. The crystallized samples were polished to the thickness up to a half of the diameter, and took optical microscope photos, scanning electron microscope (SEM) image (FEI, Quanta 200 FEG), X-ray powder diffraction (XRPD, Rigaku, Rad-C model), and microwave dielectrics    $_r$ and $Qf$ by Hakki and Coleman method[10,11]. The crystalline phase ratios between indialite and cordierite were determined by Rietveld method Fllprof software[12] using XRPD data which was taken by step scanning 0.02° for 10 sec/step by monochromatized Cu$K\alpha$ radiation generated by an X-ray tube 40 KV 20 mA. Covalency was calculated based on the bond length derived from the coordinates presented by Broun et al.[13,14]

RESULTS AND DISCUSSION

XRPD patterns of indialite/cordierite glass ceramics crystallized at 1200, 1300 and 1400 °C for 10 h are refined using indialite and cordierite mixed models by Rietveld method. Fig. 1 shows one of the XRPD patterns, which is crystallized at 1400 °C for 10 h. The coordinates of indialite and cordierite are shown in Table 1(a) and 1(b), respectively, which are refined based on the ICSD (36248) and ICSD

(30947 & 86344), respectively[15]. The site occupancies of cordierite are adopted from previous paper[16], which are ordered on Al and Si. As the determination of the site occupancies is very difficult because of the similarity of the atomic scattering factors of Al and Si, we will estimate the values from the size of polyhedra volumes and covalency as describe later. On the other hand, in the case of indialite, the site occupancies of Al and Si are disordered. Their patterns of those crystallized at 1200, 1300 and 1400 °C are well fitted as the weighted reliability factors $R_{wp}$ 2.50, 2.60 and 2.13 % for indialite and 4.79, 2.86 and 2.47 % for cordierite, respectively. The amounts of indialite and cordierite are shown in Table 2. Amount of indialite is large of 96.7 % at low temperature of 1200 °C. On the other hand, amount of cordierite of 82.9 % is large at high temperature of 1400 °C.

Figures 2 shows amount of indialite phase (a), quality factor $Qf$ values, dielectric constants $_r$, and temperature coefficient of resonant frequency $TCf$ (b) on the cordierite/indialite glass ceramics crystallized at the range of 1200 ~ 1440 °C for 10 h.-The $Qf$ values shows 148,000 GHz at 1320 °C, which is more improved value than the highest $Qf$ value of 90,000 GHz obtained by adding Ni by using the conventional solid state reaction[16]. The $Qf$ values measured in this series are decreasing to 70,000 GHz with increasing temperature to 1440 °C. The percentage of indialite phase in the total crystalline phase (cordierite + indialite) at 1200 °C for 10 h is 96.7 % and decreases to 17.1 % at 1400 °C. These values for contents of indialite and cordierite are renewed against previous paper[4], because the coordinates of the both phases are refined more precisely than those of previous ones. Based on the relationship between precipitated phases and the $Qf$ values, the sample including indialite as major constituent shows high $Q$. Indialite having the high symmetry due to the disordered crystal structure has higher $Q$ than cordierite having the low symmetry due to ordered crystal structure.

The dielectric constants $_r$ of all samples are about 4.7 as shown in Fig. 2(b). This is excellent property for millimeter-wave dielectrics, because low dielectric constant is required for time delay shortage and accuracy on manufacturing. Temperature coefficients of resonate frequency $TCf$ of the samples crystallized at 1200~1440 °C for 10 h are around -26 to -29 ppm/°C as shown in Fig.2(b). These values are also excellent for millimeter-wave applications due to near zero $TCf$.

But the measured $Qf$ values showed a wide scattering range, which are dependent on the sintering conditions in every samples. Most of the samples have clacks as shown in the preparing paper[3], which are generated by anisotropic crystal growth initiated from the surface. Even though indialite/cordierite has small thermal expansion coefficient, cracks are produced by the difference of the coefficient between [100]$_h$ and [001]$_h$. Here, [100]$_h$ means the direction [100] on the hexagonal lattice. The cracks produced due to the anisotropic crystal growth affects to the $Qf$ values. The scattering range of $Qf$ values also should be considered from crystallization conditions such as crystallinity, density, and grain size as presented by Koga et al.[6]. But it was very difficult to find the differences in these parameters among the samples. In this glass ceramics case, it is difficult to observe the microstructure because of different grain growth behavior contrasting with ceramic case. Grains nucleated from the sample surfaces grow directionally into the central region as shown in the preparation paper[4]. Fig. 3 shows surfaces of as grown samples crystallized at 1200 °C for 10 h. The grain sizes are around 30 to 100 m. The grain size is dependent on the number of seeds on the surface. The grains on the surface are growing from the crystalline seeds, and making boundaries with other grains on the surface like a ceramics. However, the number of the nucleated grain on the surfaces would be the same regardless of the crystallization time. The grains grow from the surface to inside and elongate to about 3 mm from the surface to make anisotropic grains. So, grain growth of such anisotropic large size grains is difficult.

As the determination of the site occupancy of Al and Si is difficult as described before, volumes of $AlO_4$ and $SiO_4$ tetrahedra are compared for the tendency of ordering of Al and Si. In our previous paper concerning with Ni-doped cordierite[16-18], we presented one knowledge that the crystal structure changed from cordierite to indialite was shown by change of volume of $Al/SiO_4$ tetrahedra and covalency of Al/Si ions. Fig. 4 shows volumes of tetrahedra as a function of crystallization temperature. In the case of crystal structure of indialite, there are two sites for Al and Si denoted as $Al(1)_{IND}$ and $Si(1)_{IND}$. These sites are occupied disorderly by Al and Si: $Al(1)_{IND}$ of 70% Al and 30% Si, and $Si(1)_{IND}$ of 70% Si and 30% Al[15]. These site occupancies changes as crystallization temperature. The volume of $Al(1)_{IND}$ is increasing from 2.4 to 2.6 $Å^3$ as shown in Fig. 4. The increasing shows increasing of amount of Al ions in Al(1) site. On the other hand, that of $Si(1)_{IND}$ is decreasing from 2.2 to 2.1 $Å^3$. This means increasing of amount of Si ions in $Si(1)_{IND}$ site. This tendency is showing that the positions of tetrahedra are changing disorder to order which means changing from indialite with disorder structure to cordierite with order structure. In the case of cordierite, volumes of $Al(1)_{COD}$ and $Al(2)_{COD}$ mainly occupied by Al ions increase from around 2.4 A3 to 2.6 $Å^3$ according to amount of Al ions with large ionic radius, and volumes of $Si(1)_{COD}$, $Si(2)_{COD}$ and $Si(3)_{COD}$ mainly occupied by Si ions decrease from 2.35 to 2.2 $Å^3$ according to amount of Si ions with small ionic radius. Fig. 5 shows covalencies[13,14] of Al and Si ions as a function of crystallization temperature. In the case of crystal structure of indialite, covalency of $Si(1)_{IND}$ ion increases and that of $Al(1)_{IND}$ ion decreases as increasing crystallization temperature 1200 from to 1400 °C. This tendency shows ordering of Al and Si as increasing temperature. In the case of cordierite, the covalencies of $Si(1)_{COD}$, $Si(2)_{COD}$ and $Si(3)_{COD}$ increase, and that of $Al(1)_{COD}$ and $Al(2)_{COD}$ decrease depending on the amount of Al and Si ions occupancy, because the covalency of Si ion is larger than that of Al ion. Here, these volumes and the covalencies of cordierite are obtained from the coordinate of single phase model by Rietveld analysis, though those values of indialite are obtained from two-phases mixture model.

The data obtained in this paper could be added for an example of high $Q$ due to high symmetry. On the compounds with order-disorder transition, high $Q$ is brought by high symmetry instead of ordering. These results show that high symmetry is the most important feature, though ordering is important one for high $Q$. This conclusion has been derived from following five examples: (1) in the case of $Ba(Zn_{1/3}Ta_{2/3})O_3$ (BZT), the $Q$ values are not depended on the ordering ratio[5,6]. (2) In the case of $Ba(Zn_{1/3}Nb_{2/3})O_3$ (BZN)[8] with an clear order-disorder transition, the low temperature form with high ordering ratio did not show higher $Q$ than the high temperature form with high symmetry. (3) The disordered BZT samples synthesized by spark plasma sintering (SPS) showed high $Q$ the same as ordered ones synthesized by solid state reaction (SSR)[7]. (4) Ni-doped cordierite[16-18] changing to disordered high temperature form was improved in the $Q$ value. (5) In the case of tungstenbronze without order-disorder transition, compositional ordering brings high $Q^{19}$.

CONCLUSIONS

Indialite/cordierite glass ceramics are fabricated by crystallization of glass melted at 1550 °C. Indialite is one of polytypes of cordierite, which is an intermediate phase of crystallization from glass to indialite. The amount of indialite is 96.68 % at low crystallization temperature of 1200 °C for 10 h, and decreased to 17.10 % at 1400 °C. The $Qf$ value showed the tendency of decreasing from 140,000 to 70,000 GHz, though the values are fluctuated by clacking during anisotropic crystallization. It was clarified that indialite with disordered structure on Al and Si tertrahedra shows higher $Q$ than cordierite with ordered structure. This indialite/cordierite glass ceramics has been presented as one of examples that high symmetry brings high $Q$ instead of ordering on the compound with order-disorder transition. And also this glass ceramics is expected for millimeter-wave dielectrics because of high $Q$ and low $_r$.

ACKNOWLEDGMENTS

The authors thank to Dr. Chunting Lee for the measurement of millimeter-wave dielectric properties and Professors Ken'ichi Kakimoto for supporting research.

REFERENCES

1H. Ohsato, "Microwave materials with high $Q$ and low dielectric constant for wireless communications," *Mater. Res. Soc. Symp. Proc.*, **833**, 55-62 (2005).

2Fine Ceramics Center (JFCC): Report of support industrial project 2011, "R&D for ceramics substrates with low thermal expansion and high thermal conductivity"

3H. Ohsato, in preparation for publication.

4H. Ohsato, J-S. Kim, A-Y. Kim, C-I. Cheon, and K-W. Chae, "Millimeter-wave Dielectric Properties of Cordierite/Indialite Glass Ceramics," *Jpn. J. Appl. Phys.*, 50 09NF01 1-5 (2011).

5E. Koga and H. Moriwake, "Effects of Superlattice Ordering and Ceramic Microstructure on the Microwave $Q$ Factor of Complex Perovskite-Type Oxide Ba(Zn$_{1/3}$Ta$_{2/3}$)O$_3$," *J. Ceram. Soc. Jpn,* **111** 767-775 (2003) [in Japanese].

6.E. Koga, Y. Yamagishi, H. Moriwake, K. Kakimoto and H. Ohsato, "Large $Q$ factor variation within dense, highly ordered Ba(Zn$_{1/3}$Ta$_{2/3}$)O$_3$ system," *J. Euro. Ceram. Soc.*, **26** 1961-1964 (2006).

7E. Koga, H. Moriwake, K. Kakimoto, and H. Ohsato, "Synthesis of Disordered Ba(Zn$_{1/3}$Ta$_{2/3}$)O$_3$ by Spark Plasma Sintering and Its Microwave $Q$ Factor," *Jpn. J. Appl. Phys.*, **45**(9B) 7484-7488 (2006).

8E. Koga, Y. Yamagishi, H. Moriwake, K. Kakimoto, and H. Ohsato, "Order-disorder transition and its effect on Microwave quality factor $Q$ in Ba(Zn$_{1/3}$Nb$_{2/3}$)O$_3$ system," *J. Electroceram*, **17** 375-379 (2006).

9H. Ohsato, E. Koga, I. Kagomiya, and K. Kakimoto, "Origin of High $Q$ for Microwave Complex Perovskite," *Key Engineering Materials*, **421-422** 77-80 (2010).

10B.W. Hakki and P.D. Coleman, "A Dielectric Resonator Method of Measuring Inductive in the Millimeter Range," *IRE Trans. Microwave Theory & Tech.* **8** 402-410 (1960).

11Y. Kobayashi and M. Kato, *IEEE Trans. Microwave Theory & Tech.*, **33** 586-592 (1985).

12Fullprof software by Juan Rodriguez-Carvajal in France, http://www-llb.cea.fr/fullweb/powder.htm.

13I. D. Brown and R. D. Shannon, "Empirical Bond-Strength-Bond-Length Curves for Oxides," *Acta. Cryst.*, **A29** 266-282 (1973).

14I. D. Brown and K-K. Wu, Empirical Parameters for Calculating Cation-Oxygen Bond Valences. *Acta. Cryst.*, **B32**, 1957-1959 (1976).

15ICSD

16M. Terada, K. Kawamura, I. Kagomiya, K. Kakimoto, and H. Ohsato, "Effect on Ni substitution on the microwave dielectric properties of cordierite," *J. Euro. Ceram. Soc.*, **27**(3), 3045-3148 (2007).

17H. Ohsato, I. Kagomiya, M. Terada, and K. Kakimoto, "Origin of improvement of $Q$ based on high symmetry accompanying Si-Al disordering in cordierite millimeterwave ceramics," *J. Eur. Ceram. Soc.*, **30** 315-318 (2009).

18H. Ohsato, M. Terada, I. Kagomiya, K. Kawamura, K. Kakimoto, and E. S. Kim, "Sintering Condition of Cordierite for Microwave/Millimeterwave Dielectrics," *IEEE Trans. Ultrason., Ferroelect., Freq. Contr.*, **55**(5) 1081-1085 (2008).

19H. Ohsato, "Science of tungstenbronze-type like $Ba_{6-3x}R_{8+2x}Ti_{18}O_{54}$ ($R$= rare-earth) microwave dielectric solid solutions," *J. Euro. Ceram. Soc.*, **21** 2703-2711 (2001).

Table 1. The coordinates of indialite (a) and cordierite (b) obtained by Rietveld analysis.

Table 1(a)

| Coordinate of indialite for 1200 °C 10 h (WP7) | | | | | | |
|---|---|---|---|---|---|---|
| Name | Atom | Position Multiplicity Wyckoff letter | Occupancy | $x$ | $y$ | $z$ |
| Mg | Mg | 4c | 0.167 | 1/3 | 2/3 | 1/4 |
| Al(1) | Al | 6f | 0.175 | 1/2 | 1/2 | 1/4 |
|  | Si |  | 0.075 |  |  |  |
| Si(1) | Si | 4b | 0.350 | 0.3718(2) | 0.2658(2) | 0 |
|  | Al |  | 0.150 |  |  |  |
| O(1) | O | 24m | 0.5 | 0.4857(2) | 0.3486(2) | 0.1443(2) |
| O(2) | O | 12l | 0.5 | 0.2314(3) | 0.3094(3) | 0 |

Table 1(b)

| Coordinate of cordierite for 1400 °C 10 h (WP9) | | | | | |
|---|---|---|---|---|---|
| Name | Position Multiplicity Wyckoff letter | Occupancy | $x$ | $y$ | $z$ |
| Mg | 8g | 0.5 | 1/3 | 0 | 1/4 |
| Al(1) | 8k | 0.5 | 1/4 | 1/4 | 0.279(4) |
| Al(2) | 8l | 0.5 | 0.0511(2) | 0.3100(4) | 0 |
| Si(1) | 4b | 0.25 | 0 | 1/2 | 1/4 |
| Si(2) | 8l | 0.5 | 0.1918(2) | 0.0782(4) | 0 |
| Si(3) | 8l | 0.5 | 0.1353(2) | 0.7615(4) | 0 |
| O(1) | 8l | 0.5 | 0.0426(4) | 0.7453(8) | 0 |
| O(2) | 8l | 0.5 | 0.1226(4) | 0.1889(7) | 0 |
| O(3) | 8l | 0.5 | 0.1634(4) | -0.0803(7) | 0 |
| O(4) | 16m | 1.0 | 0.2460(3) | -0.1005(4) | 0.3575(6) |
| O(5) | 16m | 1.0 | -0.0638(3) | 0.4168(5) | 0.3501(7) |
| O(6) | 16m | 1.0 | -0.1742(3) | -0.3136(4) | 0.3547(6) |

Table 2. Amount of indialite/cordierite of glass powders crystallized at 1200 to 1400 °C for 10 h, which are analyzed by Rietveld method.

| Crystallization temperature oC | Indialite % | Cordierite % |
|---|---|---|
| 1200 | 96.68 | 3.32 |
| 1300 | 53.86 | 46.19 |
| 1400 | 17.10 | 82.90 |

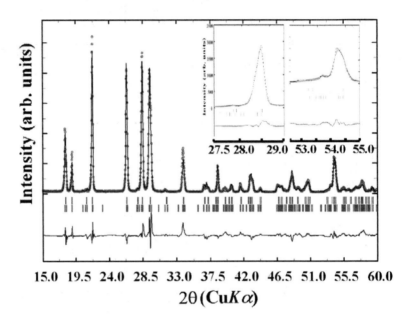

Fig. 1 XRPD patterns of glass powder crystallized at 1400 °C for 10 h, analyzed by Rietveld method using two-phase model. Differences between indialite and cordierite are shown in the inset figures.

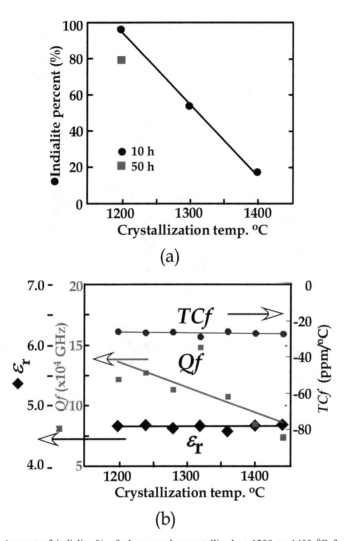

Fig. 2. (a) Amount of indialite % of glass powder crystallized at 1200 to 1400 °C for 10 h. (b) Millimeter-wave dielectric properties $Qf$, $\varepsilon_r$, and $TCf$ of glass ceramics crystallized at 1200 to 1440 °C.

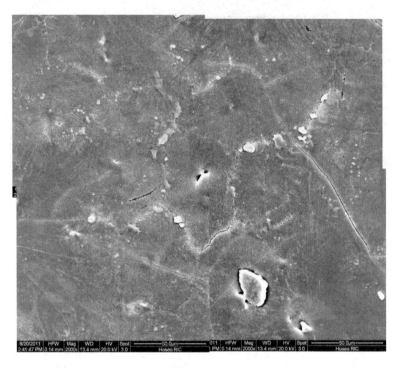

Fig. 3. Grains on the surface of glass ceramics crystallized at 1200 oC for 10 h. Crystals grow radially from nucleus on the center of the grains.

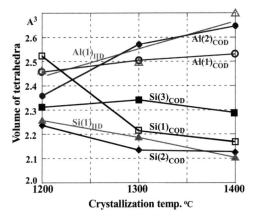

Fig. 4. Volumes of Al/SiO4 tetrahedra on the glass ceramics crystallized at 1200 to 1400 °C for 10 h, which are calculated from coordinates obtained by Rietveld analysis.

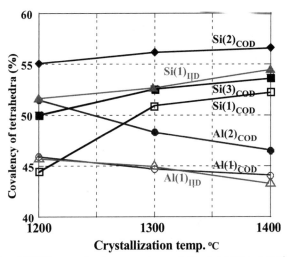

Fig. 5. Covalencies of Al/Si ions on the glass ceramics crystallized at 1200 to 1400 °C for 10 h, which are calculated from bond lengths obtained by Rietveld analysis.

# DIELECTRIC PROPERTIES OF NB-RICH POTASSIUM LITHIUM TANTALATE NIOBATE SINGLE CRYSTALS

Jun Li[1, 2,*], Yang Li[1, 2], Zhongxiang Zhou[2], Ruyan Guo[1], Amar Bhalla[1]

[1]Multifunctional Electronic Materials and Device Research Lab
Department of Electrical and Computer Engineering
The University of Texas at San Antonio, San Antonio 78249, USA

[2]Department of Physics, Harbin Institute of Technology
Harbin 150001, China

ABSTRACT

High quality $(K_{0.95}Li_{0.05})(Ta_{1-x}Nb_x)O_3$ (KLTN), $x$=0.44 ~ 0.66, single crystals have been successfully grown by the top-se eded melt growth m ethod. Based on the dielectric property measurements, three phase trans ition temperatures are determined. The Curie tem peratures can be adjusted from 328 K to 478 K with increas ing niobium content in the KL TN compositions. Low dielectric loss values are typically found at near room temperatures. Between heating and cooling runs, a hys teretic temperature-dependent dielectric behavior is noticed in all crys tals measured. The good crystal quality obtained and th e desirable room temperature dielectric properties of the KLTN crystals reported make this family of crystal material highly attractive for further investigations.

## INTRODUCTION

Most of the reports about dielectric properties of $KTa_{1-x}Nb_xO_3$ (KTN) crystals are available so far for the Ta-rich compositions. $KTaO_3$ is an incip ient ferroelectric (quantum paraelectric) material below about 13K [1] and the Curie tem perature of ferroelectric $KNbO_3$ is about 700 K [2]. The cubic lattice parameters of $KTaO_3$ is a~4.0026 Å while that of $KNbO_3$ is a~4.0226 Å, which are very close in value [3], and they make a good solid solution KTN system. According to the phase diagram, KTN com positions have four dif ferent crystallographic states: from rhombohedral, orthorhombic, tetragonal to cubic phases with the increase of temperature. The Curie temperature can be tailored in the range of 13K to 700K by adjusting the T a/Nb ratio. In KTN s ystem, the Curie temperature obeys $T_c$=276$(x-0.008)^{1/2}$ K [4] for $x$<0.05, follows the relation $T_c$=676$x$+32 K [5] for $x$>0.35, and shows weak diffuse phase transition characteristics for $x$ is in between.

However, Nb-rich KTN com positions are dif ficult to grow. To facilitate a good single crys tal growth, 5% of Li on A-site is introd uced to grow high quality Nb-rich $(K_{0.95}Li_{0.05})(Ta_{1-x}Nb_x)O_3$ (KLTN) single crystals. In this paper we report th e dielectric properties of a series of Nb-rich KLTN single crystals.

## EXPERIMENTAL

High purity starting materials K $_2CO_3$, $Li_2CO_3$, $Ta_2O_5$ and Nb $_2O_5$ powders were prepared according to the com position selected in the phase diag ram.[6] The top-seeded m elt growth

---

* Corresponding author: jun.li2@utsa.edu

method was used to grow KLTN single crystals. During the growth, the rotation rate of the seed rod was set to 10 rpm , and the pulling speed was 0.5 mm /h. After finishing the growth process, the annealing step was followed. The rate of temperature change for annealing and cooling w as kept very low (at about 20 °C/h). After 6~7 days of full growth run, high quality single crystals of KLTN were obtained.

After cutting, polishing, and stre ss releasing (thermal annealing), the crystal samples with different sizes and aspect ratios desirab le for various experimentations were prepared. The dielectric properties were m easured as a functi on of tem perature and at different frequencies using an HP4284 LCR meter, through a programmed data acquisition with automatic control for experimental variables.

RESULTS AND DISCUSSION

(1) KLTN single crystals

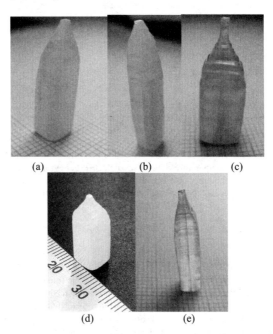

(a)          (b)          (c)

(d)          (e)

**Figure 1.** The photos of $K_{0.95}Li_{0.05}Ta_{1-x}Nb_xO_3$ single crystals grown by the Top-seeded melt growth method: (a) $x$=0.44, (b) $x$=0.50, (c) $x$=0.56, (d) $x$=0.60, (e) $x$=0.66.

Figure 1 shows the photos of KLTN single crystals of vari ous compositions grown by the top-seeded melt growth method. The compositions correspond to $x$=0.44, 0.50, 0.56, 0.60, 0.66, respectively. Because the KLTN single crystal compositions are in the ferroelectric state at room temperature, the high density of 90 ° degree domains make the crystal less transparent, but the optical transmission after polishing, poling etc. were up to the    desirable optical transparency level required for various optical applications.

(2) Dielectric behaviors

$KTa_{1-x}Nb_xO_3$ (x>0.35) single crystals under   go rhombohedral, orthorhombic, tetragonal ferroelectric states and to cubic paraelec tric state at the tran sition temperature, $T_C$, while being heated from low to high tem   perature up to  ~500 K. The trans   ition temperature from rhombohedral to orthorhombic phase is designated as  $T_{R-O}$, orthorhombic to tetragonal as  $T_{O-T}$ and finally tetragonal to cubic as $T_C$.

Figure 2 ~ 6 show the dielectric behaviors of    KLTN single crystals of unpoled samples in the heating and cooling runs.

Both the dielectric constant and loss tangent for  x=0.43 sample are shown in Figure 2 (a) & (b) as solid and dashed lines, respectively . The corresponding transition temperatures, $T_C$ ≈330 K and 325 K, $T_{R-O}$ 80 K a nd 75 K in the heating and coolin g run respectively, are observed. The typical dielectric constant a nd loss tangent at room   temperature are approximately 6,000 and 0.08 respectively; and the m aximum values ~ 13,000 and 0.60 were observed for the KL   TN crystals with x=0.43.

Table 1 su mmarizes the tran sition temperature of various com positions as derived from   the dielectric constant and loss tang ent vs. temperature data. With increasing niobium content, the Curie temperature increases as expected.

**Table 1.** Dielectric constant, loss tangent and transition temperatures of $K_{0.95}Li_{0.05}Ta_{1-x}Nb_xO_3$ single crystals: (a) in the heating run; (b) in the cooling run

(a)

| x | $T_{R-O}(K)$ | $T_{O-T}(K)$ | $T_C(K)$ | $\varepsilon_{rmax}$ | $\tan\delta_{max}$ | $\varepsilon_{RT}$ | $\tan\delta_{RT}$ |
|---|---|---|---|---|---|---|---|
| 0.44 | 80 | -- | 329 | 11, 200 . | 0.626 | 6, 390 | 0.080 |
| 0.50 | 86 | -- | 371 | 6, 230 | 0.984 | 3, 910 | 0.582 |
| 0.56 | 70 | -- | 420 | 12, 600 | 1.390 | 6, 470 | 0.428 |
| 0.60 | -- | -- | 442 | 3, 690 | 0.510 | 586 | 0.509 |
| 0.66 | 120 | 260 | 478 | 3, 380 | 0.291 | 858 | 0.083 |

(b)

| x | $T_{R-O}(K)$ | $T_{O-T}(K)$ | $T_C(K)$ | $\varepsilon_{rmax}$ | $\tan\delta_{max}$ | $\varepsilon_{RT}$ | $\tan\delta_{RT}$ |
|---|---|---|---|---|---|---|---|
| 0.44 | 75 | -- | 325 | 13, 100 | 0.576 | 7, 270 | 0.074 |
| 0.50 | 78 | -- | 357 | 6, 510 | 0.984 | 4, 320 | 0.582 |
| 0.56 | 60 | -- | 410 | 13, 100 | 1.300 | 6, 370 | 0.326 |
| 0.60 | 130 | -- | 423 | 3, 880 | 0.483 | 407 | 0.475 |
| 0.66 | 112 | 246 | 496 | 3, 470 | 0.303 | 1, 060 | 0.229 |

(3) Temperature hysteresis

The dielectric measurements show temperature hysteresis of the transitions between heating and cooling runs (at the dielectric p eak values) as shown in figure 7,  suggesting the first-order type phase transition. Generally, the tran sition temperatures derived in the heating runs are h igher than the values in the cooling runs for such firs t order behavior, but in the current studies for the crystals with $x$=0.66, the same behavior was observed in ferroelectric phase transitions; however the $T_C$ has a reverse trend. The $T_{R-O}$ and $T_{O-T}$ in the heating runs are 120 K and 260 K respectively, and in the cooli ng run those values co rrespond to 1 12 K and 246 K respectively . The Curie temperature $T_C$ is found to be ~ 478 K in the hea ting run that is lower than 496 K as observed in the cooling run. Such anom alous behavior of x=0.66 crystals needs further investigations.

CONCLUSION

Li-doped $KTa_{1-x}Nb_xO_3$, i.e. $(K_{0.95}Li_{0.05})$ $(Ta_{1-x}Nb_x)$ $O_3$, with $x$=0.44 to 0.66 single crystals have been successfully grown using the top-seeded melt growth m ethod. The crystals are of good optical quality. Dielectric m easurements reveal three dielectric anom alies corresponding to the three phase transitions presented in all the cry stals of these compositions within the temperature range measured. Also for crystals with $x$=0.44 ~ 0.66, the Curie tem peratures increases with the increase of niobium ($x$). The temperature hysteresis of the dielectric peaks corresponding to the first order phase transition of various transi tions has been noted and m easured. Several measurements of the electrical and optical pro perties are in progress and the results will be reported in the forthcoming publications.

ACKNOWLEDGEMENTS:

This work has been supported by the China Sc holarship Council, the US National Science Foundation, INAMM grant #0844081; the V alero funds at the UTSA, and National Natural Science Foundation of China grant #11074059.

REFERENCE

1.  J. K. Hulm, B. T. Matthias, and E. A. Long. A Ferromagnetic Curie Point in KTaO₃ at Very Low Temperatures. *Physical Review*. 1950, **79**:885

2.  B. T. Matthias and J. P. Remeika. Dielectric Properties of Sodium and Potassium Niobates. *Physical Review*. 1951, **82**: 727

3.  S. Triebwasser. Study of Ferroeletric Transitions of Solid-Solution Single Crystals of KNbO₃-KTaO₃. *Physical Review*, 1959, **114**: 1

4.  U. T. Höchli, H. E. Weibel, L. A. Boatner. Quantum Limit of Ferroelectric Phase Transitions in KTa₁₋ₓNbₓO₃. *Physical Review Letter*. 1977, **39**: 1158

5.  C. H. Perry, R. R. Hayes, N. E. Tornberg, in *Molecular Spectroscopy of Dense Phases*, edited by M. Goosman, S. G. Elkomoss, J. Ringeisen (Elsevier, Amsterdam, 1976), p. 267

6.  A. Reisman, S. Treibwasser, F. Holtzberg. *Journal of the American Chemical Society*. 1955, **77**: 4228

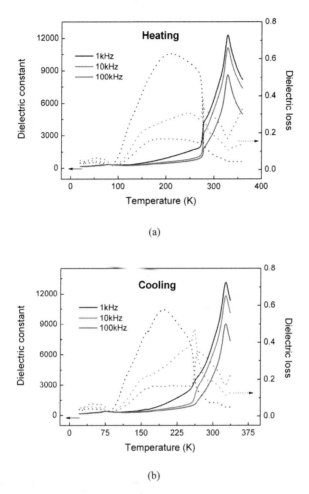

(a)

(b)

**Figure 2.** Dielectric constant and loss of $x=0.44$ sample as a function of temperature at different frequencies: (a) dielectric constant and loss on heating, (b) dielectric constant and loss on cooling.

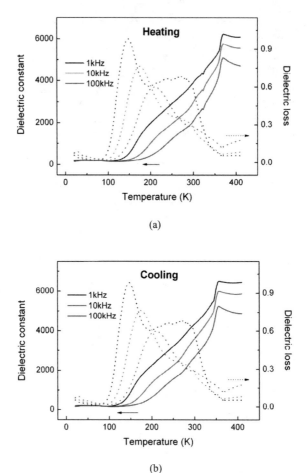

(a)

(b)

**Figure 3.** Dielectric constant and loss of $x$=0.50 sample as a function of temperature at different frequencies: (a) dielectric constant and loss on heating, (b) dielectric constant and loss on cooling.

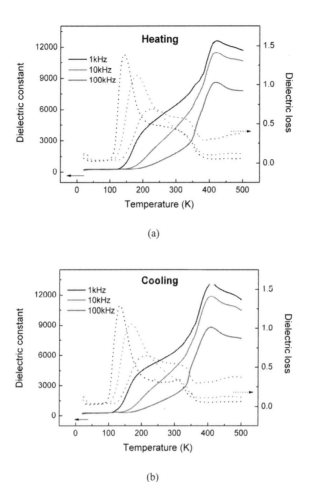

(a)

(b)

**Figure 4.** Dielectric constant and loss of $x$=0.56 sample as a function of temperature at different frequencies: (a) dielectric constant and loss on heating, (b) dielectric constant and loss on cooling.

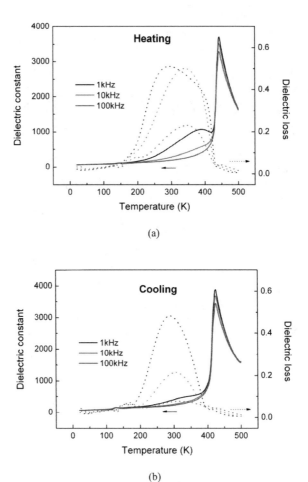

(a)

(b)

**Figure 5.** Dielectric constant and loss of $x$=0.60 sample as a function of temperature at different frequencies: (a) dielectric constant and loss on heating, (b) dielectric constant and loss on cooling.

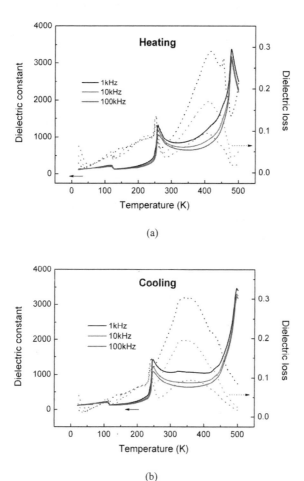

(a)

(b)

**Figure 6.** Dielectric constant and loss of $x$=0.66 sample as a function of temperature at different frequencies: (a) dielectric constant and loss on heating, (b) dielectric constant and loss on cooling.

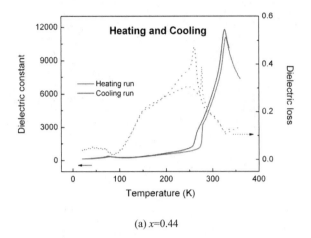

(a) $x$=0.44

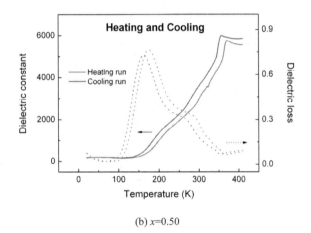

(b) $x$=0.50

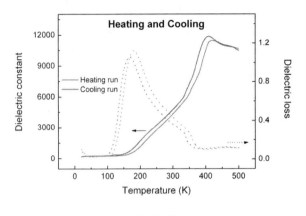

(c) $x$=0.56

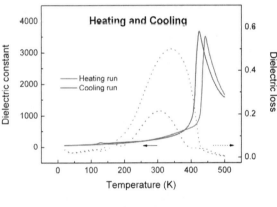

(d) $x$=0.60

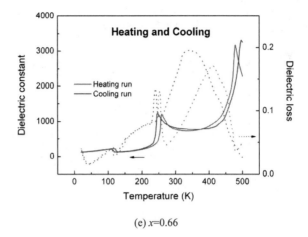

(e) $x$=0.66

**Figure 7.** Dielectric constant and loss of $K_{0.95}Li_{0.05}Ta_{1-x}Nb_xO_3$ single crystals in the heating and cooling runs at 10 kHz: (a) $x$=0.44, (b) $x$=0.50, (c) $x$=0.56, (d) $x$=0.60, and (e) $x$=0.66.

# ELECTRICAL PROPERTIES OF CALCIUM TITANATE:HYDROXYAPATITE COMPOSITES

Madhuparna Pal*, A.K. Dubey†, B.Basu†, R.Guo*, A.Bhalla*

*Department of Electrical and Computer Engineering, The University of Texas at San Antonio
One UTSA Circle, San Antonio, Texas 78249

†Indian Institute of Technology Kanpur, Kanpur-208016, India

ABSTRACT

Perovskite $BaTiO_3$:hydroxyapatite (HAp) composites have shown promises for possible applications in synthetic bone materials. This paper reports the synthesis and electrical properties of composites of the perovskite family calcium titanate, a well known ferroelastic material that shows quantum paraelectric-like dielectric behavior at temperatures below 50K, with hydroxyapatite. The dielectric and pyroelectric properties of these composites have been studied. The results show that $CaTiO_3$ retains its characteristic behavior in the composite with HAp and thus its quantum paraelectric-like behavior is sustained. Results are discussed in view of the standard Barrett equation for the quantum paraelectric or incipient ferroelectric behavior.

## INTRODUCTION

Composites of ferroelectric $BaTiO_3$ and Hydroxyapatite $[Ca_{10}(PO_4)_6(OH)_2]$ have recently been studied for their potential applications as bone analog material [1]. As an extension of this work, composites of calcium titanate (CTO) and hydroxyapatite have been synthesized and studied for their suitability as a potential material for orthopedic implant applications. The preliminary data on these composites is very promising; the advantages of CTO-HAp composite over HAp have been discussed in details [2]. In this work, we have measured and evaluated the dielectric and pyroelectric properties of these composites.

At room temperature, CTO has orthorhombic structure with lattice parameters a = 5.367 Å, b = 7.644Å and c = 5.444Å. The dielectric constant as a function of temperature behaves more like a quantum paraelectric or incipient ferroelectric materials like $SrTiO_3$ and $KTaO_3$. In a typical quantum paraelectric (QPE) case, the dielectric constant does not vary with temperature below certain temperature. These temperatures as reported in the literature are ~ 30K for $SrTiO_3$ and ~5K for $KTaO_3$[3-4].

In this regime, the ferroelectric fluctuations (and polarization) in the paraelectric phase are overtaken by the quantum fluctuations. The dielectric properties of a quantum paraelectric phase can be well described by the Barrett equation, $\varepsilon = A + \dfrac{C}{(T_1/2)\coth(T_1/2T) - T_0}$, where A is the temperature independent part of the dielectric constant, C is the Curie-Weiss constant, $T_1$ is the saturation temperature which is related to the quantum mechanical zero point motion of the elementary dipoles, $T_0$ is the extrapolated Curie temperature of the ferroelectric transition and T is the temperature [5]. Below the leveling off temperature, the dielectric constant $\varepsilon$ is found to be independent of the temperature. While quantum paraelectric effect has been known in $SrTiO_3$ and $KTaO_3$, $CaTiO_3$ was reported to have similar quantum paraelectric like dielectric constant vs. temperature behavior [6]. CTO is considered as a new member of the family of quantum paraelectric and research has been focused on the role of chemical substitutions in suppressing the quantum fluctuations. The dielectric constant ε for CTO increases continuously as the temperature is decreased from 300 to 12 K. The leveling-off temperature for CTO has been suggested in various previous reports to be ~ 35 K [7].

**Table 1.** Comparison of the fitting parameters of CTO with some classic QPE like STO and KTaO

| | $\varepsilon(RT)$ | $\varepsilon_{max}$ | $A$ | $C, 10^4$ K | $T_0$, K | $T_1$, K | Reference |
|---|---|---|---|---|---|---|---|
| CaTiO₃ | 168 | 331 | 43.9 | 4.77 | −111 | 110 | [7] |
| SrTiO₃ | 305 | 20 000 | | 8 | 35.5 | 80 | [3] |
| KTaO₃ | 239 | 3840 | 47.5 | 5.45 | 13.1 | 56.9 | [4] |

Hydroxyapatite is an important biomaterial for applications related to bone and dental transplants. The earliest use of HAp as an implant was made in the alveolar bone in 1977 [8]. It is bone compatible and binds to the bone tightly without preventing connective tissue encapsulation. The ceramic HAp possesses poor mechanical and electrical properties. One of the approaches is to combine ceramic HAp with a ferroic material in order to increase its electrical activity. BaTiO₃ and CaTiO₃, belonging to the same perovskite family and being bio-friendly materials, have been tried for the synthesis of HAp:BTO and HAp:CTO composites. The CTO has proven to be a good substrate for the apatite growth since it provides a positive surface charge that interacts with the negatively charged phosphate ions in the fluid [9]. This paper focuses on the measurements of dielectric and pyroelectric current generation characteristics of the synthesized composites as a function of temperature.

EXPERIMENTAL

The composites were prepared using different concentrations of CTO (80%, 60% and 40% by weight) and with HAp. These powders were mixed homogeneously, shaped and sintered into disc of diameter 5-6 mm and thickness of 1-2 mm. Details of sample preparation are given in our earlier reports [9]. Dielectric and pyroelectric measurements were conducted on these samples. The variation of ε with temperature was measured as a function of frequency. The temperature was varied from 323K to 15K in order to include the leveling temperature well in the range. The measurements were repeated for five different frequencies (100 Hz, 1 kHz, 10 kHz, 100 kHz and 1 MHz). Attempts were also made to study the pyroelectric current as a function of temperature.

The composites of CTO and HAp were gold plated using the sputter coater. The sputtering was carried out at a chamber pressure of 0.05 mtorr, while the sputter current was maintained at around 20 mA. Silver wires were attached to the gold plated surface using silver paste. The silver wires were used for making all electrical connections. The capacitance and the tangent loss (tan δ) measurements were done with the HP 4284A LCR meter, which has a range from 20 Hz to 1 MHz. The temperature controller was integrated into the experimental setup for temperature variation. The GADD[1] software was used for data collection.

---

[1] a computer program developed by Paul Moses for electric multispectroscopy measurement.

RESULTS AND DISCUSSION

Dielectric data collected on various composites of compositions i.e. with 80%, 60% and 40% (weight %) CTO are shown in figures 1(a), 1(b) and 1(c) respectively. It is interesting to note that, in each case, the dielectric constant as a function of temperature increases while cooling and finally levels off at around 40-50 K range. In appearance, the behavior resembles to that of a typical QPE like STO and KTaO$_3$.

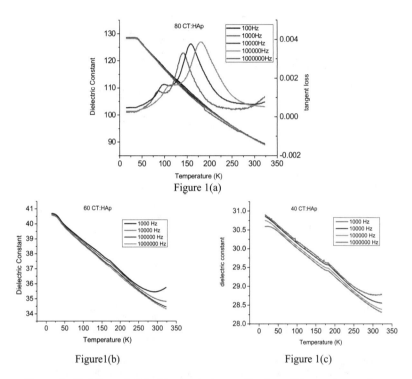

Figure 1(a)

Figure1(b)                Figure 1(c)

**Figure 1.** Variation of dielectric constant as a function of temperature and frequency in various composites of CTO: HAp (a) 80% CT (b) 60% CT and (c) 40% CT. Figure 1(a) also shows the tangent loss.

Dielectric plots show the closeness to the QPE behavior and thus the Barrett equation was used to analyze the results. Figures 2(a) and 2(b) show the fitted curve along with the dielectric data collected at 1 kHz frequency. The corresponding fitting parameters are listed in Table 2. It should be noted that only the 80% and 60% compositions could be fitted well with the Barrett equation. Though the nature of the dielectric constant for the 40% composition shows a resemblance with the QPE like behavior, as shown in the figure 2(c), it could not be fitted very well as the volume percentage of CaTiO₃ in the composite might have been reduced below the critical point. The interesting part of these plots and measurements is the integrity of CTO in the synthesized composites. Such well fitted dielectric data to the Barrett equation excludes the possibility of apparent chemical reactivity of CTO and HAp which otherwise would have influenced the QPE like dielectric behavior of CTO.

**Table 2**. The fitting parameters for 80% and 60% CT:HAp composites

| Model | Barrett equation | |
|---|---|---|
| Equation | $y = A + C/(T_1/(2*tanh(T_1/(2*x)))-T_0)$ | |
| Composition | 80 CT:HAp | 60 CT:HAp |
| A | 33.577 | 28.330 |
| C | 35093.818 | 4612.253 |
| $T_1$ | 123.476 | 15.058 |
| $T_0$ | -306.664 | -380.646 |

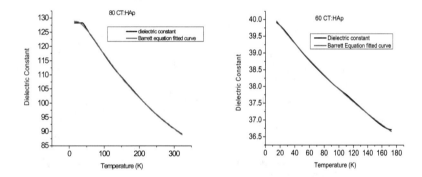

**Figure 2** (a) and (b). Dielectric constant and corresponding fitting curves with the Barrett equation for (a) 80 CT:HAp and (b) 60CT:HAp, respectively, at 1 kHz. suggesting the QPE nature of the CTO: HAp composite.

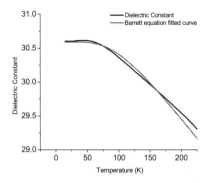

**Figure 2(c).**
Dielectric constant as a function of temperature at 1 MHz for 40% CT:HAp. The curve resembles the quantum paraelectric-like behavior but does not entirely fit with the Barrett equation.

Pyroelectric measurements are performed on the poled samples. Each sample was cooled from 375 K to 15 K under the bias electric field of 10 kV/cm. After the field was removed at 15 K, the sample was shorted for 10-30 minutes before collecting the depolarized current in the heating cycle from 15 to 375 K. In each case a maxima in the depolarization current was observed around 350 K. The magnitude of the maxima was different for different composition samples and there were some effects due to the different level of applied electric fields used for the poling purpose. The figures 3, 4 and 5 show the depolarization current with the temperature for 80%, 60% and 40% CTO (weight %) compositions respectively.

Once these samples have gone through the depolarizing process in the heating cycle, the depolarization current maxima could not be reproduced either in the cooling cycle or in the second heating cycle. The property could be reproduced during the second poling and heating process. The sense of the depolarization current reverses during the poling process when the direction of the applied field is reversed.

Further studies about the nature of these currents whether they are true pyroelectric or thermally stimulated or mixed in origin are desirable. Irrespective of their origin, these currents are stable, significant near the room temperature, and its maxima occur around 350 K, i.e. ~75 °C. Whether these currents and their polarity play a role in the implanted material is to be answered with further detailed studies. Some initial results in these aspects are interesting and encouraging that will be reported in the forthcoming publications.

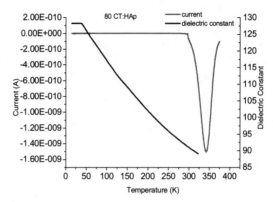

**Figure 3**. Dielectric constant and current as a function of temperature (15 K-375 K) for 80% CTO composition.

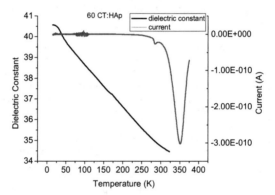

**Figure 4**. Dielectric constant and current as a function of temperature (15 K- 375 K) for 60% CTO composition.

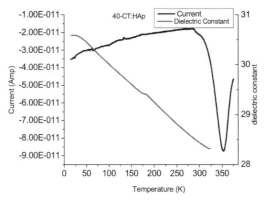

**Figure 5.** Dielectric constant and depolarization current as a function of temperature (15 K - 375 K) for 40% CTO composition.

CONCLUSION

Composites of CTO: HAp have been synthesized and studied for their dielectric and pyroelectric properties. Dielectric properties in all the composites of 80 CTO:HAp, 60CTO:HAp and 40 CTO:HAp (weight%) resemble the typical QPE-like dielectric properties of CTO. Barrett equation fits well in 80% and 60% CTO composites. The results suggest that CTO in the composites maintains its integrity and apparently does not react with HAp to form a new composition/compound. An interesting depolarization current peak in the heating run of the poled samples has been observed in all the compositions. The studies suggest that desirable composition that is most suitable as a bone replacement material can be synthesized. The dielectric and depolarization currents can be tailored according to the needs. Further experiments are needed to authenticate its origin and its role in the tissue growth experiments.

ACKNOWLEDGEMENT

The work is partially supported by NSF IMI/INAMM grant # 0844081 and by ONR grant #N00014-08-1-0854.

REFERENCES

1. Dubey, A.K., S.D.Gupta, and B. Basu, *Optimization of electrical stimulation parameters for enhanced cell proliferation on biomaterial surfaces.* Journal of Biomedical Materials Research Part B: Applied Biomaterials, 2011. **98B** (1): p. 18-29.
2. Dubey, A.K., G.Tripathi, and B. Basu, *Characterization of hydroxyapatite-perovskite (CaTiO₃) composites: Phase evaluation and cellular response.* Journal of Biomedical Materials Research Part B: Applied Biomaterials, 2010. **95B** (2): p. 320-329.
3. Müller, K.A. and H. Burkard, *SrTiO₃: An intrinsic quantum paraelectric below 4 K.* Physical Review B, 1979. **19** (7): p. 3593-3602.

4. Samara, G.A. and B. Morosin, *Anharmonic Effects in KTaO₃: Ferroelectric Mode, Thermal Expansion,and Compressibility.* Physical Review B, 1973. **8** (3): p. 1256-1264.
5. Barrett, J.H., *Dielectric Constant in Perovskite Type Crystals.* Physical Review, 1952. **86** (1): p. 118-120.
6. Ang, C., A.S. Bhalla, and L.E. Cross, *Dielectric behavior of paraelectric KTaO₃, CaTiO₃, and Ln₁/₂Na₁/₂TiO₃ under a dc electric field.* Physical Review B, 2001. **64**(18): p. 184104-110.
7. Lemanov, V., *Perovskite CaTiO₃ as an incipient ferroelectric.* Solid State Communications, 1999. **110** (11): p. 611-614.
8. Ducheyne, P. and K. De Groot, *In vivo surface activity of a hydroxyapatite alveolar bone substitute.* Journal of Biomedical Materials Research, 1981. **15** (3): p. 441-445.
9. J. Coren, O. Coren, Evaluation of calcium titanate as apatite growth promoter. Journal of Biomedical Materials Research Part A, 2005. **75A** (2):p. 478-484.

THE INFLUENCE OF CONSOLIDATION PARAMETERS ON GRAIN CONTACT SURFACES BaTiO₃-CERAMICS

Vojislav V. Mitic[1,2], Vladimir B. Pavlovic[2,3], Vesna Paunovic[1], Miroslav Miljkovic[1], Jelena Nedin[1], Milan Dukic[4]

[1]Faculty of Electronic Engineering, University of Nis, Nis, Serbia
[2]Institute of Technical Sciences of SASA, Belgrade, Serbia
[3]FoA, University of Belgrade, Serbia
[4]North Carolina Central University, Durham, USA

ABSTRACT
    Among the modified BaTiO₃ La doped BaTiO3-ceramics are intensively studied because of their attractive ferroelectric properties that can be tailored to meet the strong demands for various electroceramic components. In this investigation La/Mn codoped BaTiO₃ with different La₂O₃ content, ranging from 0.1 to 5.0 at% La, was investigated regarding their microstructural characteristics. Microstructural and compositional studies were performed by scanning electron microscopy (JEOL-JSM 5300) equipped with EDS (QX 2000S) system. Depending on the amounts of dopants and sintering temperature a huge range of microstructure features can be observed in doped ceramics, together with the appearances of secondary phases that are formed above eutectic temperature. The morphology of ceramics grains pointed out the validity of developing new structure analytical methods, based on different grains' shape geometries. The new correlation between microstructure of doped BaTiO₃-ceramics, based on fractal geometry and contact surface probability, has been developed. Using the fractals and statistics of the grains contact surface, a reconstruction of microstructure configurations, as grains shapes or intergranular contacts, has been successfully done. The directions of possible materials properties prognosis are determined according to the correlations synthesis-structure-property.

INTRODUCTION
    The one of the most investigated ferroelectric materials nowadays are BaTiO₃-based ceramics due to their broad applications in advanced electronics. It is well known that intergranular structure and electric properties of these materials can be controlled by using different technological parameters and different additives. Slight change of particular consolidation parameter, such as pressing pressure, initial sample's density, sintering temperature and time, or the change of dopant concentration, can significantly change microstructure and final properties of electronic materials [1-4]. Among the modified BaTiO₃ compositions, La and La/Mn codoped BaTiO₃ ceramics are intensively studied because their attractive dielectric response can be used in various electronic devices. The incorporation of La³⁺, which replaces A sites in perovskite BaTiO₃ structure, modifies the microstructural and electrical properties of doped BaTiO₃. The substitution of La³⁺ on Ba²⁺ sites requires the formation of negatively charged defects. There are three possible compensation mechanisms: barium vacancies ($V_{Ba}''$), titanium vacancies ($V_{Ti}''''$) and electrons ($e'$). For samples sintered in air atmosphere, which are the electrical insulators, the principal doping mechanism is the ionic compensation mechanism. In order to reduce the dissipation factor, MnO or MnO₂ are frequently added to BaTiO₃ together with other additives. It has been established that for donor-acceptor La/Mn-codoped BaTiO₃ ceramics, with small grained microstructure, the formation of donor-acceptor complexes such as 2[$La_{Ba}^{•}$]-[$Mn_{Ti}^{//}$] prevent a valence change of Mn²⁺ to Mn³⁺.
    Beside dopant concentration it has been established that grain size distribution may also considerably affect dielectrical properties of BaTiO₃ based materials. Therefore correlation between

their microstructure and dielectrical properties has been investigated by numerous authors [5-8]. The grainy structure of BaTiO₃-ceramics is so fairly complicated that any method of classical geometry fails to describe it. The same refers to the distribution of intergrain contacts. Modeling method allows the presentation of grains in contact in the shape of sphere, ellipsoid and polyhedron.

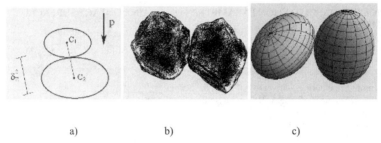

a)                    b)                    c)

Fig. 1. a) Ellipsoidal approximation, b) and c) two grains in contact according to
ellipsoid-ellipsoid model.

Recently, it has been established that since the real intergrain contact surface is an irregular object, the theory of fractal sets can be used for the reconstruction of microstructure configurations [9-12].

Fractal method traces a new approach for describing and modeling the grains shape, as well as relations among dielectrical properties and microstructure. It has been established that, modeling random microstructures, as grains aggregates in polycrystals, patterns of intergranular cracks theory of Iterated Function Systems (IFS) and the concept of Voronoi tessellation, can be used [13]. Voronoi tessellation represents cell structure constructed from a Poisson point process, by introducing planar cell walls perpendicular to lines connecting neighboring points. Considering all the above-mentioned, methods for modeling grain boundary surface and geometry of grain contacts of doped BaTiO₃-ceramics, are presented in this article.

EXPERIMENTAL PROCEDURE

In this study reagent grade BaTiO₃, (Ba/Ti = 0.996±0.004, Rhone Poulenc, average particle size of 0.10–5 μm), La₂O₃ and MnO₂ (Merck, Darmstadt), were used as starting materials. The content of La₂O₃ ranged from between 0.10 and 5.0 at% and that of MnO₂ was kept constant at 0.05 at%. The specimens are denoted such as 0.10 La-BT for specimen with 0.10 at% La and 0.05 at% Mn. The samples of La/Mn-codoped BaTiO₃ ceramics were prepared by a conventional mixed oxide solid state reaction. The starting powders were homogenized in ethanol medium with alumina balls for 24 h. After drying for several hours, the powders were pressed at 120 MPa into disks of 10 mm in diameter and 2 mm of thickness. The samples were sintered at 1320°C and 1350°C for two hours with heating rate of 300°C/h in air atmosphere. The microstructures of sintered and chemically etched samples were observed by scanning electron microscope (JEOL-JSM 5300) equipped with energy dispersive X-ray analysis spectrometer (EDS-QX 2000S system).. The illustrations of the microstructure simulation, were generated by Mathematica 6.0 software. Statistical method was used for calculation of contact surfaces.

RESULTS AND DISCUSSION

Ceramic grains contacts are essential for understanding complex electrodynamics properties of sintered materials. Microstructures of sintered BaTiO$_3$-ceramics obtained by SEM method are characteristic examples of complex shape geometry, which cannot easily be described or modeled.

Our microstructure analysis showed that the homogeneous and completely fine-grained microstructure with grain size ranged from 1.0–3.0 μm, without any indication of abnormal grains, is the main characteristics of low doped La-BaTiO$_3$, sintered at 1320°C (Fig. 2).

a)                                              b)

Fig. 2. SEM of La-doped BaTiO$_3$ sintered at 1320°C: a) 0.1 La-DT and b) 1.0 La-BT

The similar microstructure with grain size around 3.0 μm is observed in 0.1 and 0.5 at% La-BaTiO$_3$ specimens sintered at 1350°C (Fig. 3). With the increase of dopant amount the increase of porosity is evident.

a)                                              b)

Fig. 3. SEM of specimens sintered at 1350°C: a) 0.1 La-BT and b) 1.0 La-BT

EDS analysis showed that the difference in microstructural features is mainly due to the inhomogeneous distribution of La$_2$O$_3$. Taken from different area in the same sample, EDS spectra indicated that the La-rich regions are associated with small grained microstructure (Fig. 4a), while the abnormal grains with domain structure are characteristic of the regions which were free of La-content (Fig. 4b).

a)                                                          b)

Fig. 4. EDS spectra of 1.0 at% La-BaTiO₃ sintered at 1350°C: a) local area reach in La associated with fine grained matrix and b) domain structure free of La

In this article the correlation between microstructure and dielectric properties of doped BaTiO₃-ceramics, based on fractal geometry and contact surface probability, has been developed. Using the fractals and statistics of the grains contact surface, a reconstruction of microstructure configurations, as grains shapes or intergranular contacts, has been successfully done.

Considering shapes properties, method of Voronoi tessellation has also been used for grains surface modeling. It is important to emphasize, that although polycrystal modeling generally represents a 3D problem, a 2D Voronoi tessellation approximation is used. This assumption is based on the fact that the sample properties mostly depend on microstructure surfaces.

Set of points is generated in order to represent the distribution of micro-grains. In this case, the distribution is semi-random and implemented in such manner that each point dislocates in relation to its position on the imaginary lattice. The main idea is to establish correlation between the different points-peaks and micro values between two contact surfaces, which practically represent the new level, more complex and realistic, micro intergranular capacitors-impendances network (Fig. 5).

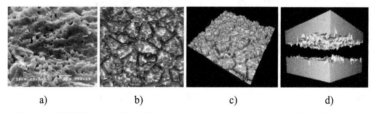

a)                    b)                    c)                    d)

Fig. 5. a) SEM image of BaTiO₃ ceramics and b), c), d) corresponding 3D microsurface overview.

Irregularity of the surface of ceramics grains can be expressed using the term called fractal dimension. Determination of grain contour fractal dimension includes several steps. First of all, coordinates of chosen points from the contour must be determined in relation to some rectangular coordinate system. Coordinates $X$ and $Y$ are linear dimensions. Then, the method of iterative functional systems (IFS) is applied, in the form of fractal interpolation, modified to parameter level. Method of fractal interpolation can be improved by taking contour points coordinates (sampling) in few different positions of the same contour. The contour of the grain obtained by fractal interpolation is given in Fig. 6. Resulting fractal dimension is estimated to range from 1.118 to 1.164.

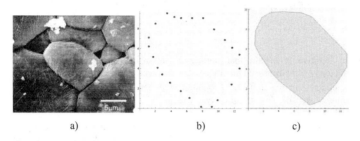

a)                                    b)                                    c)

Fig. 6. a) SEM microphotograph, b) set of chosen points from the contour and c) fractal interpolation method for segregate grain of BaTiO₃ ceramics.

Due to diffusional forces that appear in sintering process, it is supposed that an approximate form of contact surface is the shape of a minimal surface – the surface with minimal area size. But, the microstructure of the material makes this surface locally fractal (Fig. 7a). Considering that intergrain contact surface is a region, where processes occur at the electronic level in the electroceramic material, structural complex grain contact-grain can be represented by an electrical equivalent network.

Contact between two grains is observed as planar microcapacitor (Fig. 7b). The surfaces of capacitor plates correspond to intersection surface (Sc) of two grains.

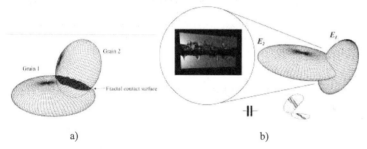

a)                                    b)

Fig. 7. a) An intergrain contact surface has fractal form, b) two grains in contact form a microcapacitor.

Therefore, the formula for the capacitance of the planar capacitor, formed in the contact, is given by

$$C = \varepsilon_0 \varepsilon_B \frac{S_c}{x}$$

(7)

where $\varepsilon_0$, $\varepsilon_B$ are dielectric constants in vacuum and BaTiO₃-ceramic material, respectively; $Sc$ - the area of the plates and $x$ - distance between capacitor plates, i.e. the capacitor thickness.

In fractal case, according to fractal approach to the intergrain geometry, the formula (7) is modified by introducing correction factor $\alpha$, $\alpha = (N\xi^2)^k$. Thus, the formula for the microcapacity of intergranular capacitor, seen as planar capacitor, is given in the following form

$$C = \varepsilon_0 \varepsilon_B \alpha \cdot \frac{S}{x} = \varepsilon_0 \varepsilon_B (N\xi^2)^k \cdot \frac{S}{x} \qquad (8)$$

This approach uses an iterative algorithm that iterates $N$ self-affine mappings with constant contractive (Lipschitz) factor $|\xi| < 1$ $k$-times. Typically, $\alpha = D - D_T$, where $D$ is the fractal (Hausdorff) dimension of intergrain contact surface and $D_T$ is topological dimension of the surface. BaTiO$_3$-ceramics contact surfaces are of low-irregularity, which is characterized by the small difference $D - D_T \approx 0.08744$.

Derived formula (8) indicates lower values of microcapacity, when fractal approach is applied. Thus, more accurate calculation of microcapacitance, generated in grains contact, can be carried out, leading to a more exact estimation of dielectric properties of the whole sample.

CONCLUSIONS

In this article, using the fractals and statistics of the grains contact surface, a reconstruction of microstructure configurations, as grains shapes or intergranular contacts, has been successfully done. For better and deeper characterization and understanding of the ceramics material microstructure, the methods that include the fractal nature structure, and also Voronoi model and mathematical statistics calculations, are applied. Voronoi is one specific interface between fractal structure nature and different stochastical contact surfaces, defined by statistical mathematical methods. Also, the Voronoi model practically provided possibility to control the ceramics microstructure fractal nature. Mathematical statistic methods enabled establishing the real model for the prognosis of correlation: synthesis-structures-properties.

Obtained results indicated that fractal analysis and statistics model for contact surfaces of different shapes leads to better understanding of the influence of the fractal nature on final microstructure and dielectrical properties of BaTiO$_3$-ceramics.

ACKNOWLEDGEMENTS

*This research is part of the project "Directed synthesis, structure and properties of multifunctional materials" (172057) and of the project III44006. The authors gratefully acknowledge the financial support of Serbian Ministry of Education and Science for this work.*

REFERENCES

[1] S. Wang, G. O. Dayton, Dielectric Properties of Fine-Grained Barium Titanate Based X7R Materials, J. Am. Ceram. Soc., 82 (10), (1999), 2677–2682.

[2] C. Pithan, D. Hennings, R. Waser, Progress in the Synthesis of Nanocrystalline BaTiO$_3$ Powders for MLCC, International Journal of Applied Ceramic Technology, 2 (1), (2005), 1–14.

[3] Y.K. Cho, S.L.Kang, D.Y. Yoon, Dependence of Grain Growth and Grain-Boundary Structure on the Ba/Ti Ratio in BaTiO$_3$, J. Am. Ceram. Soc., 87, (2004), 119-124.

[4] P. Kumar, S. Singh, J.K. Juneja, C. Prakash, K.K. Raina, Influence of calcium substitution on structural and electrical properties of substituted barium titanate, Ceramics International, 37, (2011), 1697–1700.

[5] R.Zhang, J.F.Li, D.Viehland, Effect of aliovalent supstituents on the ferroelectric properties of modified barium titanate ceramics: relaxor ferroelectric behavior, J. Am. Ceram. Soc., 87 (5), (2004), 864-870.

[6] Z.C.Li, B.Bergman, Electrical properties and ageing characteristics of BaTiO$_3$ ceramics doped by single dopants, J. E. Ceram. Soc., 25, (2005), 441-445.

[7] P. W. Rehrig, S. Park, S. Trolier-McKinstry, G. L. Messing, B. Jones, T. Shrout, Piezoelectric properties of zirconium-doped barium titanate single crystals grown by templated grain growth, J. Appl. Phys., 86 (3), (1999), 1657-1661.

[8] V. P. Pavlovic, M. V. Nikolic, V. B. Pavlovic, N. Labus, Lj. Zivkovic, B. D. Stojanovic, Correlation between densification rate and microstructure evolution of mechanically activated BaTiO$_3$, Ferroelectrics, 319, (2005), 75-85.

[9] V. V. Mitic, V. B. Pavlovic, Lj. Kocic, V. Paunovic, D. Mancic, Application of the Intergranular Impedance Model in Correlating Microstructure and Electrical Properties of Doped BaTiO$_3$, Science of sintering, 41 (3), (2009), 247-256.

[10] W. Klonowski, E. Olejarczyk, R. Stepien, A new simple fractal method for nanomaterials science and nanosensors, Materials Science Poland, 23(3), (2005), 607-612.

[11] V. Mitic, V. Pavlovic, Lj. Kocic, V. Paunovic, Lj.Zivkovic, Fractal geometry and properties of doped BaTiO$_3$ ceramics, Advances in Sciency and Technology, 67, (2010) 42-48.

[12] V. V. Mitic, Z. S. Nikolic, V. B. Pavlovic, V. V. Paunovic, D. Mancic, B. Jordovic, Lj. M. Zivkovic, Influence of Yb$_2$O$_3$ and Er$_2$O$_3$ on BaTiO$_3$ ceramics microstructure and corresponding electrical properties, Developments in Strategic Materials, 29 (10), (2008), 231-236.

[13] V. Mitic, V. Paunovic, D. Mancic, Lj. Kocic, Lj. Zivkovic, V.B. Pavlovic, Dielectric Properties of BaTiO$_3$ Doped with Er$_2$O$_3$, Yb$_2$O$_3$ Based on Intergranular Contacts Model, Advances in Electroceramic Materials, Ceramic Transactions, 204, (2009), 137-144.

# Magnetoelectric Multiferroic Thin Films and Multilayers

FERROIC AND STRUCTURAL INVESTIGATIONS IN RARE EARTH MODIFIED TbMnO₃ CERAMICS

G. S. Dias, R. A. M. Gotardo, I. A. Santos, L. F. Cótica
Grupo de Desenvolvimento de Dispositivos Multifuncionais, Departamento de Física, Universidade Estadual de Maringá
Maringá - Paraná, Brazil. 87020-900

J. A. Eiras, D. Garcia
Grupo de Cerâmicas Ferroelétricas, Departamento de Física, Universidade Federal de São Carlos
São Carlos, São Paulo, Brazil. 13565-905

ABSTRACT
In this work, high-dense TbMnO₃ ceramics were processed from nanocrystalline powders obtained through high-energy ball milling. The structure and microstructure of dense ceramics were studied by X-ray diffraction and scanning electron microscopy, and revealed high-dense and almost single-phased samples. The dielectric properties were investigated and showed two relaxation processes in the temperature range of 100 K to 300 K, and a third phenomenon at ~75 K that could be related to well-known magnetic transitions of TbMnO₃.

INTRODUCTION
In the recent years, the interest for investigating lead-free materials as a possible alternative to widely used lead-based ceramics has been growing. TbMnO₃ have attracted great attention in the field of multiferroic magnetoelectric materials, since this material show the coexistence of magnetic and electric orders and, some times, a coupling between them. Although most magnetoelectric materials present only a low coupling between their magnetic and electric properties[2,3], the rare earth manganites, RMnO₃, where R = Gd, Tb and Dy, have emerged as a new class of ferroic materials with strongly coupled (anti)ferromagnetic and ferroelectric properties[3]. In these materials, specially in TbMnO₃, ferroelectricity occurs at cryogenic temperatures mainly due to a transition between two magnetic states[3,4]. In fact, TbMnO₃ single crystals present an incommensurate sinusoidal antiferromagnetic ordering at $T_N$ = 41 K, and the magnetic wavenumber $k$ is incommensurate at $T_N$. $k$ decreases with the decrease of the temperature until be locked to a constant value ($T_{lock-in}$ = 27 K), where a polarization arises in the $c$ – axes direction[4]. Interesting, this polarization flops into the $a$ direction when a magnetic field (higher than 4 T) is applied along the $a$ or $b$ directions, while the polarization is suppressed in the $c$ direction[5,6].

EXPERIMENTAL
The Tb₀.₉La₀.₁MnO₃ samples were synthesized from analytical grade terbium, manganese, and lanthanum oxides (Alfa Aesar). The starting powders were milled in a Retsch PM 100 planetary ball mill at 400 rpm for 12 h, and pre-fired at 1473 K in air for 1 h. Before sintering, the pre-fired powders were milled again for 48 h in air. These powders were uniaxially pressed into disks, and isostatically pressed at 148 MPa. These samples were then sintered at 1673 K, in oxygen atmosphere, for 5 h. The samples' structure was investigated by X-ray diffraction in a Shimadzu XRD-7000 diffractometer. The microstructures were revealed using a Shimadzu Super Scan SS-550 scanning electron microscope. Some disk surfaces were gold coated, by sputtering, for dielectric characterizations that were carried out in a JANIS CCS-400H/204 cryostat, using the standard four-probe method with a high precision Agilent E4980A LCR bridge.

RESULTS AND DISCUSSION

Figures 1(a) and (b) show the scanning electron microscopy analyses for the $Tb_{0.9}La_{0.1}MnO_3$ sample milled for 12 h and pre-fired for 1 h at 1473 K in air. This sample exhibits the starting stage of sintering process showing large particles agglomerates, higher than 100 μm. This kind of microstructure is not favorable for obtaining high densified samples. This way, the pre-fired powders were milled again for 48 h for eliminating large agglomerates and reduce the particle sizes. The SEM results for re-milled powders are shown in Figures 1(c) and (d), revealing very small particle sizes with a narrow particle size distribution, centered on 300 nm.

Figure 1. Scanning electron microscope images for $Tb_{0.9}La_{0.1}MnO_3$ samples. (a) and (b) pre-fired for 1 h at 1473 K in air; (c) and (d) Pre-fired powders milled for 4 h at 400 rpm.

Non-agglomerated powders were used to process ceramic bodies, which were sintered in oxygen atmosphere to get charge balance control. Figures 2(a) and (b) show the scanning electron microscope images that unambiguously reveal high-dense samples that are largely cracked (as a result of an intense shrinkage) and present a microstructure completely absent of large pores. An image of a fractured ceramic surface (Figure 2(b)) attests the absence of large pores and reveals predominantly transgranular fractures as the main feature of the well-sintered ceramic body.

Figure 2. Scanning electron microscope images for the $Tb_{0.9}La_{0.1}MnO_3$ sample sintered at 1673 K for 5 h in oxygen atmosphere. (a) Surface and (b) fracture.

The Rietveld refinement results and the X-ray diffraction patterns, for the $Tb_{0.9}La_{0.1}MnO_3$ powdered ceramic, are shown in Figure 3. The X-ray diffraction analysis revealed the presence of two distinct phases. One of them, the majority phase, was identified as showing a structure similar to those of $TbMnO_3$, exhibiting an orthorhombic symmetric unit cell. The cell parameters are listed in Table 1. The second phase was identified as being the $TbO_2$ cubic phase, which represents about 3% of the sample batch, and emerged due to the very high sintering temperature used for sample densification. Furthermore, the substitution of Tb by La caused a small increase of the unit cell volume in comparison with pure $TbMnO_3$ (see Table 1). This unit cell enlargement can be attributed to the difference between Tb and La ionic radius, which is higher for lanthanum.

Table 1 Cell parameters for sintered $Tb_xLa_{1-x}MnO_3$ samples.

| | a Å | b Å | c Å | Volume Å³ |
|---|---|---|---|---|
| $TbMnO_3$ | 5.3140 | 5.8146 | 7.4330 | 229.201 |
| $Tb_{0.9}La_{0.1}MnO_3$ | 5.2998 | 5.8417 | 7.4031 | 229.673 |

Figures 4(a) and (b) show the temperature dependence of the real ($\varepsilon'$) complex permittivity and loss tangent (tg $\delta = \varepsilon''/\varepsilon'$, where $\varepsilon''$ is the imaginary part of the complex permittivity) for $Tb_{0.9}La_{0.1}MnO_3$ ceramic. It can be seen that $\varepsilon'(T)$ exhibits a frequency behavior similar as those reported by Wang et al.[7]. Nearby room temperature, $\varepsilon'(T)$ reaches values as high as $\sim10^4$ and, with decreasing temperature, $\varepsilon'(T)$ displays a steplike decrease to a lower value of about 15. Below 60 K, $\varepsilon'(T)$ is almost frequency and temperature independent, unlike pure $TbMnO_3$, which becomes frequency and temperature independent slightly below 100 K. This difference can be attributed to the Tb ions substitution by La.

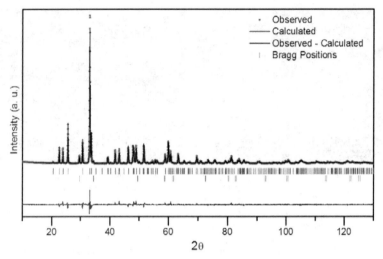

Figure 3: X-Ray diffraction patterns and Rietveld refinement results for the Tb$_{0.9}$La$_{0.1}$MnO$_3$ sample at room temperature.

The step-like decrease in $\varepsilon'(T)$ consists of two distinct phenomenon that correspond to two distinct peaks in loss tangent with peak positions shifted to higher temperatures as the measuring frequency increases. These results suggest two distinct thermally activated relaxation phenomena in Tb$_{0.9}$La$_{0.1}$MnO$_3$ ceramic. The activation energy (E) and the pre-exponential factor ($\tau_0$) for the high and low-temperature relaxations were determined as being 0.20 eV and 1.21 X 10$^{-9}$ s, and 0.13 eV and 8.28 X 10$^{-11}$ s, respectively. The activation energy exhibited by low-temperature relaxation is in complete agreement with previous results reported in the literature, i. e., 0.1467 eV[8]. On the other hand, the activation energy (0.20 eV) exhibited by the high-temperature relaxation is lower than previous results, i. e., 0.3067 eV [8]. This reduction of the activation energy can be considered as a result of the formation of secondary phases (TbO$_2$) during the sintering processes, as revealed in the X-ray diffraction analysis. Furthermore, this lowering of the activation energy certainly contributes to enhance the electrical conductivity and dissipation losses of the processed sample, as in fact revealed by the tg $\delta$ curves. The relaxation peaks in loss tangent are almost indistinct, and as reported by Wang et al.[7], the exponential increasing of the loss tangent background implies hopping conductivity associated to them.

Interesting, a third relaxation phenomena is observed in loss tangent around 75 K. However, the range of temperatures where it occurs prevented a complete characterization of this relaxation phenomenon. In fact, as this phenomenon was observed in the range of temperatures where ferroelectric and magnetic transitions takes places in TbMnO$_3$ single crystals[9], additional studies at lower temperatures are necessary to determine the true nature of this relaxation phenomenon.

REFERENCES

[1]Y. Cui, Decrease of Loss in Dielectric Properties of TbMnO$_3$ by Adding TiO$_2$, *Physica B*, Vol 403, 2008, p 2963-2966.

[2]N.A. Hill, Why Are There so Few Magnetic Ferroelectrics?, *J. Phys. Chem. B*, Vol 104, 2000, p 6694-6709.

[3]M. Fiebig, Revival of the Magnetoelectric Effect, *J. Phys. D: Appl. Phys.*, Vol 38, 2005, R123-R152.

[4]T. Kimura, T. Goto, H. Shintani, K. Ishizaka, T. Arima, and Y. Tokura, Magnetic Control of Ferroelectric Polarization, *Nature*, Vol 426, 2003, p 55-58.

[5]N. Aliouane, K. Schmalzl, D. Senff, A. Maljuk, K. Prokes, M. Braden and D. N. Argyriou, Flop of Electric Polarization Driven by the Flop of the Mn Spin Cycloid in Multiferroic TbMnO$_3$, *Phys. Rev. Lett.*, Vol 102 (No. 207205), 2009.

[6]S.B. Wilkins, T.R. Forrest, T.A. W. Beale, S.R. Bland, H.C. Walker, D. Mannix, F. Yakhou, D. Prabhakaran, A.T. Boothroyd, J.P. Hill, P.D. Hatton, and D.F. McMorrow, Nature of the Magnetic Order and Origin of Induced Ferroelectricity in TbMnO$_3$, *Phys. Rev. Lett.*, Vol 103 (No. 207602), 2009.

[7]C. C. Wang, Y.M. Cui, and L.W. Zhang, Dieletric Properties of TbMnO$_3$ Ceramics, *Appl. Phys. Lett.*, Vol 90 (No. 012904), 2007.

[8]Y. Cui, L. Zhang, G. Xie, and R. Wang, Magnetic and Transport and Dielectric Properties of Polycrystalline TbMnO$_3$, *Solid State Comm.*, Vol 138, 2006, p 481-484.

[9]T. Kimura, G. Lawes, T. Goto, Y. Tokura, and A.P. Ramirez, Magnetoelectric Phase Diagrams of Orthorhombic RMnO$_3$ (R = Gd, Tb, and Dy), *Phys. Rev. B*, Vol 71 (No. 224425), 2005.

# HR-TEM INVESTIGATIONS IN BiFeO$_3$-PbTiO$_3$ MULTIFUNCTIONAL CERAMICS

V. F. Freitas, F. R. Estrada, G. S. Dias, L. F. Cótica, and I. A. Santos
Grupo de Desenvolvimento de Dispositivos Multifuncionais, Departamento de Física. Universidade Estadual de Maringá
Maringá – PR, Brazil. 87020-900

D. Garcia and J. A. Eiras
Grupo de Cerâmicas Ferroelétricas, Departamento de Física. Universidade Federal de São Carlos
São Carlos – SP, Brazil. 13565-905

## ABSTRACT

In this work, multifunctional $(0.6)$BiFeO$_3$-$(0.4)$PbTiO$_3$ ceramics were synthesized by high-energy ball milling. Structural and ferroelectric characterizations were performed by X-ray and electron diffraction, and P vs E measurements. The interface between tetragonal and rhombohedral phases, the morphotropic phase boundary, was observed by high-resolution transmission electron microscopy. A qualitative model is proposed to elucidate the structure/property relations in these samples. The macroscopic ferroelectric behavior of the studied samples was correlated with the strain-mediated energy increasing in the MPB region.

## INTRODUCTION

The progress for controlling the macroscopic physical properties trough closely located structural control, in atomic orders (nano-science), has been the object of countless studies[1]. The advanced electromechanical materials, used in the most recent technological advanced devices, are those that display a morphotropic phase boundary (MPB), i. e., a specific region in the phase diagram that separates two distinct isostructural symmetric phases. Particularly, in perovskite-structured materials, two structural symmetries, the rhombohedral and tetragonal ones, as those observed in Pb(Zr$_{1-x}$Ti$_x$)O$_3$ (PZT), (PMN-PT), and (1-x)BiFeO$_3$(x)PbTiO$_3$ (BFO-PT) solid solutions[1], commonly coexist and can be used for tuning some macroscopic physical properties.

The 0.6BFO-0.4PT compound is a piezoelectric[2] and multiferroic (with ferroelectric and antiferromagnetic responses) material with magnetoelectric coupling[3] at room temperature, and a MPB region. Originally, the BiFeO$_3$ (BFO) matrix possesses a perovskite-type structure with rombohedral symmetry (R3c space group). However, when PbTiO$_3$ (PT) is added into BFO, the lattice symmetry changes to tetragonal, traversing the MPB region around ~30 % of PT[3]. Thereafter, if La atoms are added in samples of the tetragonal side of the MPB in the BFO-PT compounds, a new transition occurs, backing the lattice symmetry to rhombohedral. However, other MPB region is observed in 3% La doped BFPT60/40 (BLFO-PT) samples[4]. Furthermore, the MPB region of the BFO-PT system has been recently pointed out as a great candidate for technological applications. This is because BFO-PT materials with a MPB region can be used as multifunctional materials, where those properties from the two distinct structures can be used separately or in combination in one simple and advanced technological device[1,2].

In fact, the first evidence that the electromechanical properties are improved in the MPB region was presented in middle of last century[1], where a considerable increase of the piezoelectric coefficients of PZT ceramics was reported. Recently, it was discovered monoclinic symmetries in the MPB region of the PZT system and the Cm space group was pointed as the most plausible for describing these compounds, and in these monoclinic phases, polar axis are found between [001] tetragonal and [111] rhombohedral directions[5]. In this way, the high electromechanical response observed in the MPB

215

region of the PZT system was related as being due to the monoclinic distortion[6]. Subsequently, other works related the presence of monoclinic symmetries (Cc space group) in the MPB region of PZT[7,8] and PMN-PT systems[9,10]. Finally, the monoclinic symmetries (Cm space group) was also proposed for BFO-PT samples from Rietveld refinement of XRD data and from electromechanical characterizations[3,11,12].

However, novel and more conclusive studies need to be conducted to attest that the monoclinic distortions, in the MPB region of ferroelectric materials, are responsible for the improvement in their physical properties. In this work, careful analysis in MPB region of the BFO-PT ceramics were conducted by using X-ray diffraction, electron diffraction (measured and simulated), and high-resolution transmission electron microscopy techniques. A novel mechanism for explaining the link between rhombohedral and tetragonal symmetries without the presence of monoclinic symmetries was proposed.

## EXPERIMENTS

Stoichiometric powders of the $(0.6)BiFeO_3$-$(0.4)PbTiO_3$ + 3 wt% La (BLFO-PT) composition were synthesized from analytical graded (Aldrich) $La_2O_3$, $Bi_2O_3$, $Fe_2O_3$, PbO and $TiO_2$ precursors. These powders were mechanically processed by high-energy ball milling (HEBM), with a Retsch PM100 planetary ball mill in air atmosphere, as previously reported[5]. The optimized milling conditions were: ball-to-powder mass ratio (30:1), milling speed (32 rad.s$^{-1}$) and time (1 h). Subsequently, the powders were conformed in disc shapes (10 mm in diameter and 1 mm in thickness) and submitted to cold isostatic pressing (148 MPa) and reactively sintered at 1348 K for 1 h. X-ray diffraction (XRD) analyzes were conducted with a Shimadzu XRD7000 X-ray diffractometer (Cu k$_\alpha$ radiation). The ferroelectric hysteresis loops were collected at room temperature with a Sawyer-Tower circuit at 30 Hz. HRTEM and electron diffraction analysis were conducted with a Jeol JEM 3010 URP High Resolution Transmission Electron Microscope.

## RESULTS AND DISCUSSIONS

Figure 1 shows the XRD patterns for the BLFO-PT ceramic synthesized by high-energy ball milling. The position and intensity of the diffracted peaks are very well described with the two structural symmetries patterns, the tetragonal (ICSD: 156204) and the rhombohedral (ICSD: 028626) ones, respectively. The inset in Figure 1 shows the two structural symmetries and their corresponding polarization directions. Small amount of a spurious phase was also observed in the XRD patterns, which was indexed as being $Bi_2O_3$. These results point to the existence of the MPB in the BLFO-PT sample.

Figure 2 shows two polarization loops, induced by distinct electric field levels, for the BLFO-PT ceramic. The saturation and remnant polarizations indicate the characteristic ferroelectric behavior of this sample. In fact, the piezoelectric coefficients, which attest the electrically polarized state of this compound, had already reported in a previous work[2]. The saturation polarization reached expressive values (20 μC/cm$^2$), while the remnant polarization was observed as being 12 μC/cm$^2$. In addition, the coercive field reached 22 kV/cm. These results indicate the potentialities of the BLFO-PT samples for practical applications where enhancedferroelectric and piezoelectric properties are required.

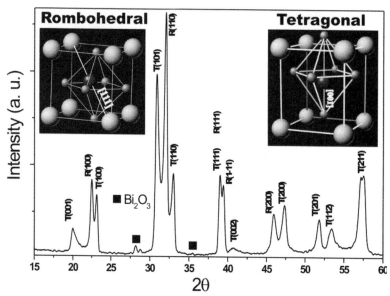

Figure 1. X-ray diffraction patterns for the (0.6)BiFeO₃-(0.4)PbTiO₃ + 3 wt% La (BLFO-PT) powdered ceramic at room temperature.

Figure 3 shows high-resolution transmission electron microscopic (HRTEM) images for the BLFO-PT sample. The electron diffraction patterns (EDP) are shown in Figure 3(a). The EDP was conducted in a particle that is presented in the low-resolution image of the Figure 3(b).

The dimensional analysis of the EDP is consistent with three families of crystallographic planes related to tetragonal, rhombohedral and cubic structural symmetries. In accordance with XRD patterns, the rhombohedral and tetragonal symmetries are related to the MPB of the BLFO-PT compound, while the cubic phase is related to the Bi₂O₃ phase. In addition, the $a_T \sim 3.85$ Å lattice parameter of the tetragonal symmetry, is very similar to the $a_R \sim 3.90$ Å lattice parameter, of the rhombohedral symmetry. A HRTEM image is showed in Figure 3(c), where an overlap of the crystalline structures is observed. This image corresponds to an amplified region of the particle marked by the frame (i) in Figure 3(b). These overlapped structures are responsible for the overlap observed in EDP analysis. Finally, the crystalline structure observed in Figure 3(d) [amplified region of the particle marked by the frame (ii), shown in Figure 3(b)] was compared with the simulated structure. The good agreement between the experimental result and the simulated results allows us state that this is the tetragonal structure, oriented in the [001] direction.

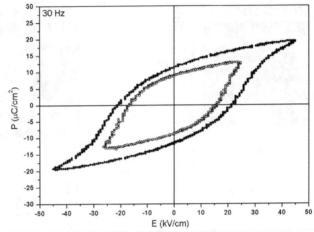

Figure 2. P vs E curves for the (0.6)BiFeO$_3$-(0.4)PbTiO$_3$ + 3 wt% La (BLFO-PT) sample at room temperature and 30 Hz.

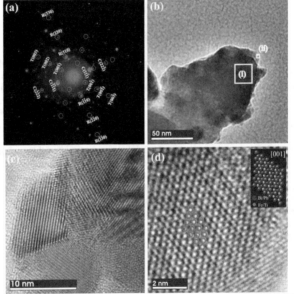

Figure 3. HRTEM analysis and electron diffraction patterns for the (0.6)BiFeO3-(0.4)PbTiO3+3wt%La (BLFO-PT) ceramic at room temperature.

Figure 4(a) shows the HRTEM image for the BLFO-PT ceramic, which represents the structural interface. This interface can be the morphotropic phase boundary, which separates the rhombohedral and tetragonal symmetries. However, this can also be the interface between two different crystallographic orientations of the same crystal. In this way, three distinct sample' regions were chosen to perform electron diffraction (inserted square frames). Figure 4(b) shows the electron diffraction pattern obtained from the frame (1), where two patterns are observed. The dimensional analysis indicates that these electron diffraction patterns are related to the tetragonal and rombohedral symmetries, as shown in Figures 4(c) and 4(d). However, the crystalline structure was not perfectly oriented with the electron been, and only one family of planes, for each symmetry, was analyzed.

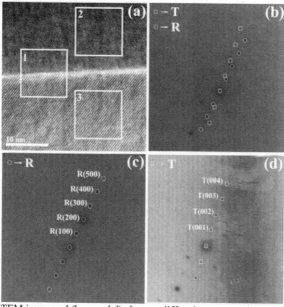

Figure 4. (a) HRTEM image and (b, c, and d) electron diffraction patterns for the atomic structure of the (0.6)BiFeO3-(0.4)PbTiO3+3wt%La (BLFO-PT) ceramic.

The physical mechanism that describes the coupling between two structural symmetries (rhombohedral and tetragonal) has been explained by the emergence of a third structural symmetry, monoclinic or orthorhombic, for linking them[5]. However, in this work, an alternative path to describe this connection, without a third symmetry between them, is proposed. In fact, a slight tilt between planes with different lattice parameters can link these two structures/symmetries. Indeed, this slight tilt can couple two different lattice parameters, such as the tetragonal $a_T$ and the rhombohedral $a_R$ (or the $c_T$ tetragonal), as shown in Figure 4(a). Therefore, the need for a third structure as a mediator of the two symmetries in the MPB region can be disregarded. In this way, this slight tilt between the two (rhombohedral and tetragonal) isostructures with different symmetries (R3c and P4mm) can be identified as a possible mechanism for explaining the enhancement of the ferroelectric properties in MPB compositions, because it gives elevated mobility to the domain walls in these samples.

CONCLUSIONS

In this work, BLFO-PT ceramics with a morphotropic phase boundary were synthesized by high-energy ball milling. Enhanced ferroelectric properties, as a consequence of the structural arrangement, were observed in the ferroelectric characterizations. The coupling between the two symmetries of the MPB region can be explained by considering a slight tilt between atomic planes of the atomic structures in both sides of the morphotropic phase boundary interface.

ACKNOWLEDGMENTS

The authors would like to thank LME/C2Nano/LNLS (proc. for experimental facilities and technical support during the electron microscopic analyses, the CNPq (proc. TEM-HR 12225) 476964/2009-1), CAPES (Procad 82/2007), and Fundação Araucária de Apoio ao Desenvolvimento Científico e Tecnológico do Paraná (Prots. 10779 and 15727) Brazilian agencies for financial support. G. S. D. (552900/2009-5) also thanks CNPq for fellowship.

REFERENCES

[1] B. Jaffe, R.S. Roth, and S. Marzullo, Piezoelectric Properties of Lead Zirconate-Lead Titanate Solid-Solution Ceramics, *J. Appl. Phys.*, **25**, 809-810 (1954).

[2] V.F. Freitas, I.A. Santos, E.R. Botero, B.M. Fraygola, D.Garcia, and J.A. Eiras, Piezoelectric Characterization of (0.6)BiFeO3–(0.4)PbTiO3 Multiferroic Ceramics, *J. Am. Ceram. Soc.*, **94**, 754-758 (2011).

[3] S. Bhattacharjee, V. Panday, R.K. Kotnala, and D. Panday, Unambiguous Evidence Form Magnetoelectric Coupling of Multiferroic Origin in $0.73BiFeO_3$-$0.27PbTiO_3$, *Appl. Phys. Lett.*, **94** 012906 (2009).

[4] V.F. Freitas, L.F. Cótica, I.A. Santos, D. Garcia, and J.A. Eiras, Synthesis and Multiferroism in Mechanically Processed $BiFeO_3$-$PbTiO_3$ Ceramics, *J. European Ceram. Soc.*, **31**, 2965-2973 (2011).

[5] B. Noheda, D. E. Cox,G. Shirane, J. A. Gonzalo, L. E. Cross, and S-E. Park, A Monoclinic Ferroelectric Phase in the $Pb(Zr_{1-x}Ti_x)O_3$ Solid Solution, *Appl. Phys. Lett.*, **74**, 2059-61 (1999).

[6] R. Guo, L. E. Cross, S-E. Park, B. Noheda, D. E. Cox, and G. Shirane, Origin of the High Piezoelectric Response in $PbZr_{1-x}Ti_xO_3$, *Phys. Rev. Lett.*, **84**, 5423-5426 (2000).

[7] B. Noheda, J.A. Gonzalo, L.E. Cross, R. Guo, S.-E. Park, D. E. Cox, and G. Shirane, Tetragonal-to-Monoclinic Phase Transition in a Ferroelectric Perovskite: The Structure of $PbZr_{0.52}Ti_{0.48}O_3$, *Phys. Rev. B*, **61**, 8687- 8695 (2000).

[8] B. Noheda, D.E. Cox, G. Shirane, R. Guo, B. Jones, and L.E. Cross, Stability of the Monoclinic Phase in the Ferroelectric Perovkite $PbZr_{1-x}Ti_xO_3$, *Phys. Rev. B*, **63**, 014103 (2000).

[9] B. Noheda, D.E. Cox, G. Shirane, S.-E. Park, L.E. Cross, and Z. Zhong, Polarization Rotation Via a Monoclinic Phase in the Piezoelectric $92\%PbZn_{1/3}Nb_{2/3}O_3$-$8\%PbTiO_3$, *Phys. Rev. Lett.*, **86**, 3891-3894 (2001).

[10] A.K. Singh, D. Panday, and O. Zaharko, Confirmations of $M_B$-type Monoclinic Phase in $Pb[(Mg_{1/3}Nb_{2/3})_{0.71}Ti_{0.29}]O_3$: A Powder Neutron Diffraction Study, *Phys. Rev. B*, **63**, 172103 (2003).

[11] S. Gupta, A. Garg, D.C. Agrawal, S. Bhattacharjee, and D. Panday, Structural Changes and Ferroelectric Properties of $BiFeO_3$-$PbTiO_3$ Thin Films Growth Via a Chemical Multilayer Deposition Method, *J. Appl. Phys.*, **105**, 014101 (2009).

[12] S. Bhattacharjee and D. Panday, Stability of the Various Crystallographic Phases of the Multiferroic $BiFeO_3$-$PbTiO_3$ System as a Function of Compositions and Temperature, *J. Appl. Phys.*, **107**, 124112 (2010).

# Multifunctional Oxides

# MODIFIED PECHINI SYNTHESIS OF La DOPED HEXAFERRITE Co₂Z WITH HIGH PERMEABILITY

Lang Qin[1]; Nahien Sharif[1]; Lanlin Zhang[2]; John Volakis[2]; Henk Verweij[1]

[1]Group Inorganic Materials Science, Department of Materials Science & Engineering, Ohio State University, 2041 N College Road, Columbus OH 43210-1178, USA

[2]ElectroScience Lab, Ohio State University, 1320 Kinnear Road, Columbus OH 43212, USA

## ABSTRACT

Z-type hexagonal ferrites with strong magnetocrystalline anisotropy are considered for applications at microwave frequencies. In particular, $Ba_3Co_2Fe_{24}O_{41}$ (Co₂Z) is expected to exhibit a high permeability and low loss at frequencies >1 GHz. In the present work, we report a systematic study on the effect of lanthanum doping on Co₂Z magnetic response. $Ba_3Co_2La_xFe_{24-x}O_{41}$ ($x$=0, 0.1, 0.5, and 1.0) was fabricated from precursor powders made by a modified-Pechini route, followed by calcination, compaction and sintering. Samples with preferred grain orientation and porosity of 75-90% were obtained. The Co₂Z magnetic properties were found to strongly depend on the microstructure. Likewise, a strong relation was found between the measured permeability and applied frequency. This is ascribed to frequency-dependant domain wall motion and spin rotation.

## INTRODUCTION

Ferrites are the most important magnetic ceramics because of their versatility in the miniaturization of microwave devices such as permanent magnets, bluetooth devices [1] and multi-layered antennas [2]. In particular, hexaferrites can enhance the magnetic field at GHz frequencies. They can be used, for instance, in ground plane antennas that can be applied on the metal surface of aircrafts and cars. Magnetic field enhancement is commonly described as soft-magnetic response. The application requirements are generally a high magnetic permeability $\mu' \geq 10$, and a low loss $\tan(\delta) \leq 0.1$, where $\tan(\delta) = \mu'/\mu''$. Hexaferrite $Co_2Ba_3Fe_{24}O_{41}$ (Co₂Z) has a high anisotropy with its crystallographic c-axis parallel to the hexagonal basal plane. Hence Co₂Z is predicted to have a high permeability $\mu' = 16$ at $v > 1$ GHz [3]. The natural ferromagnetic resonance (FMR) frequency of Co₂Z is predicted to be 3.4 GHz [3], providing a low loss $\tan(\delta)$ at frequency 3 GHz $> f > 1$ GHz. However, state-of-the-art Co₂Z has $\tan(\delta) > 0.1$ at $f > 1$ GHz, which is ascribed to the occurrence of domain wall (DW) and spin rotation (SP) resonance. The extent of domain wall motion is related to the grain sizes[4] with a typical resonance frequencies of <1 GHz [5]. Spin rotation is dependent on the anisotropy constant.

Rare earth (RE) dopants [6-8] have been reported to influence the microstructure of ferrites, such as the grain sizes, anisotropy constant, and saturation magnetization. In turn, the microstructure mayinfluence soft magnetic properties through its effects on domain motion and spin rotation. In this paper the effect of La substitution in Co₂Z on microstructure and magnetic response is studied for $Ba_3Co_2La_xFe_{24-x}O_{41}$ compositions ($x = 0, 0.1, 0.5, 1.0$).

## MATERIAL AND METHODS

Solutions were prepared by dissolving $BaCO_3$, $Co(COOCH_3)_2 \cdot 4H_2O$, $Fe(NO_3)_3 \cdot 9H_2O$ and $La(NO_3)_3 \cdot 6H_2O$ for compositios of $Ba_3Co_2La_xFe_{24-x}O_{41}$ ($x$=0, 0.1, 0.5 and 1.0) in 1 mol/L $HNO_3$ solution and citric acid. A $NH_4OH$ solution was then added to adjust the solution $p_H$ to 7.5. Subsequently these solutions were mixed with ethanol under vigorous stirring. The eventual molar ratio of metallic salt: citric acid: ethanol was 1:2:2. Most solvent was removed by heating at 300°C for 1.5 hours in an uncovered glass beaker. The obtained precursors were calcined at 600°C for 3 hours in air to remove organic residuals, using 1°C/min heating and 6°C/min cooling rates. This was followed by calcination at 1300°C for 2 hours, using 2°C/min heating and 6°C/min cooling rates. The powders were uniaxially pressed to cylindrical and toroidal specimens and sintered at 1330°C for 4 hours using 2°C/min heating and 6°C/min cooling rates.

The thermal decomposition of the citrate complex was studied with thermogravimetric analysis (TGA) in a 40 sccm air flow and 2°C/min heating up to 1350°C. The phase composition of the calcined powders and sintered samples were obtained by X-ray diffraction (XRD) using a Scintag XDS2000 with Cu Kα radiation. The micro-structure and composition distribution were examined with scanning electron microscopy (SEM) using a Quanta 200 at an accelerating voltage of 20 kV. The bulk density ($\rho^a$) of the sintered Co$_2$Z was measured using the Archimedes's principle and a Mettler Toledo Density Kit. The relative density, $\phi$, was obtained as $\phi = \rho^a/\rho^{th}$ where $\rho^{th}$ is the theoretical density of Co$_2$Z (5.35 kg/m$^3$). High frequency permeability was measured on the toroidal samples for frequencies of 10 MHz...1 GHz using an Agilent E4991A analyzer with a 16454A magnetic material test fixture.

RESULTS AND DISCUSSIONS

The four samples of Ba$_3$Co$_2$La$_x$Fe$_{24-x}$O$_{41}$ hexaferrites with $x = 0.0$, 0.1, 0.5, and 1.0 are referred to as A...D and listed in table 1. The TGA measurement revealed a complete removal of organic residual in 600°C calcined samples. Figure 1 shows the XRD patterns of the samples A...D, in which the peaks of Co$_2$Z, Ba$_2$Co$_2$Fe$_{12}$O$_{22}$ and LaFeO$_3$ are indicated. Phase-pure Co$_2$Z (ICSD 97980) was found in sample A. The four strongest peaks were assigned to, in decreasing intensity: $(00\overline{18})$, $(0116)$, $(00\overline{32})$ and $(00\overline{14})$. Theoretically for a randomly oriented sample, the strongest peak is $(0116)$. This discrepancy is ascribed to a preferential orientation with the (001) planes parallel to the top surfaces of the pellets. In the samples B, C and D, Co$_2$Y (ICSD 74487) and perovskite LaFeO$_3$ were also observed.

TABLE 1 Summary of the sintered Ba$_3$Co$_2$La$_x$Fe$_{24-x}$O$_{41}$ (x=0, 0.1, 0.5 and 1.0) samples.

| Sample | $x$ | Phase | Grain size [μm] |
|---|---|---|---|
| A | 0 | Co$_2$Z | 27 |
| B | 0.1 | Co$_2$Y, Co$_2$Z | 33 |
| C | 0.5 | Co$_2$Y, Co$_2$Z, LaFeO$_3$ | 20 |
| D | 1.0 | Co$_2$Y, Co$_2$Z, LaFeO$_3$ | 16 |

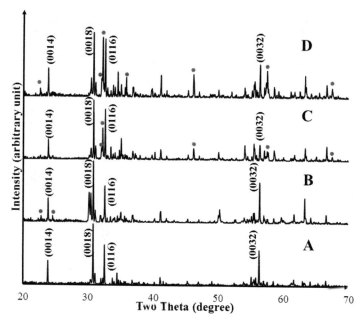

Figure 1. XRD spectra of sample A...D. Impurity Co$_2$Y and LaFeO$_3$ are labeled in "•"and "•" respectively

SEM images of samples A...D are shown in figure 2. The pressed compacts have a relative density of 70% in the initial and intermediate sintering stage. The density increases further with sintering temperature. The relative density measurements suggest that sample A has the highest relative density of 90.3%. The samples B, C and D have a relative density of 83.6%, 78.8% and 74.5%, respectively. The corresponding grain sizes are listed in table 1. The results show that the addition of La dopant leads to smaller grain sizes: 27 μm in sample A, 20 μm in sample B and 16 μm in sample C. However, the grain size increases to 33 μm in sample B, which will affect the magnetic permeability and will be explained below.

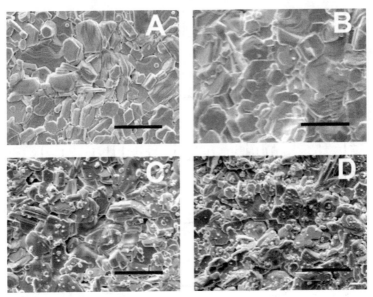

Figure 2. SEM images of sample A...D. Scale bar represents 50 μm in all images.

The permeability spectra of samples A...D are shown in figure 3. Sample A has the highest initial permeability $\mu'_A = 11.6$, followed by sample B, C and D: $\mu'_B = 9.7$, $\mu'_C = 9.1$ and $\mu'_D = 8.0$. The cutoff frequency of sample A is 500 MHz. With the increase of La-doping, the cutoff frequency decreases to 350 MHz first (sample B) and then increase to 450 MHz (sample C) and 550 MHz (sample D). The permeability of polycrystalline ferrites is related to two different magnetizing mechanisms: the spin rotational (SP) magnetization and the domain wall (DW) motion magnetization [9-11]:

$$\mu = \mu' - j\mu'' = 1 + (\mu'_{SP} + \mu'_{DW}) + j(\mu''_{SP} + j\mu''_{DW}) \qquad (1)$$

The real and imaginary parts of permeability can be described as

$$\mu'_{SP} = \frac{K_{SP}\omega_{SP}^{Res2}}{\omega_{SP}^{Res2} + \omega^2} \qquad (2)$$

$$\mu''_{SP} = \frac{K_{SP}\omega\omega_{SP}^{Res}}{\omega_{SP}^{Res2} + \omega^2} \qquad (3)$$

$$\mu'_{DW} = \frac{K_{DW}\omega_{DW}^{Res\,2}(\omega_{DW}^{Res\,2} - \omega^2)}{(\omega_{DW}^{Res\,2} - \omega^2)^2 + \beta^2\omega^2} \tag{4}$$

$$\mu''_{DW} = \frac{K_{DW}\omega_{DW}^{Res\,2}\beta\omega}{(\omega_{DW}^{Res\,2} - \omega^2)^2 + \beta^2\omega^2} \tag{5}$$

where $\omega$ is the magnetic field frequency, $K_{SP}$ is the static spin susceptibility, $\omega_{SP}^{Res}$ is the spin resonance frequency, $K_{DW}$ is the static susceptibility of domain wall motion, $\omega_{DW}^{Res}$ is the domain wall resonance frequency, and $\beta$ is the damping factor of the domain wall motion.

It is also known that $K_{SP}$ and $K_{DW}$ can be expressed as:

$$K_{SP} = \frac{4\pi M_S^2 \mu_0}{2K_1 + H_d \mu_0 M_S} \tag{6}$$

$$K_{DW} = \frac{3\pi M_S^2 D}{4\gamma} \tag{7}$$

where $H_d$ is the demagnetizing field, $K_1$ is the anisotropy constant, $M_S$ is the saturation magnetization, $D$ is the grain size and $\gamma$ is the wall energy. As shown in equation (7), $K_{DW}$ is proportional to $M_S^2 D$. The saturation magnetization $M_S$ of Co$_2$Z is 40 emu/g. In sample B, the total $M_S$ decreases due to the lower saturation magnetization of the impurity phase Co$_2$Y (28 emu/g). The grain size $D$ is increased as suggested from figure 2B, leading to a comparable $K_{DW}$ to sample A, thus the variation of $\mu''_{DW}$ and $\mu'_{DW}$ can be small. However, Co$_2$Y has a higher $K_1$ (2.6 erg/cm$^3$) than Co$_2$Z (1.8 erg/cm$^3$), resulting in a smaller $K_{SP}$ and lower $\mu''_{SP}$ and $\mu'_{SP}$. The overall results are lower loss $\mu''$ with lower permeability $\mu'$. On the other hand, sample C and D are different because of the third phase LaFeO$_3$ which has a much smaller $M_S$ of ~0.4 emu/g. Both $M_S$ and $D$ decrease with the increased La doping concentration. This will minimize the loss $\mu''_{DW}$ while decrease $\mu'_{DW}$. Additionally, the $K_1$ of LaFeO$_3$ is much smaller. The competition between $K_1$ and $M_s$ makes $K_{SP}$ adjustable by controlling the amount of La doping. As can be seen in figure 3C and D, the higher La concentrations results in lower permeability $\mu'$ and loss $\mu''$.

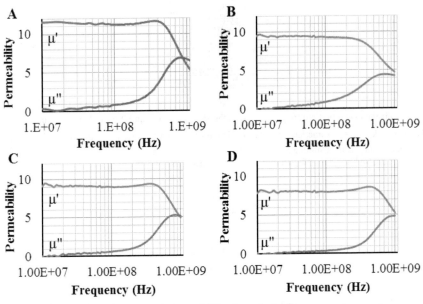

Figure 3. Frequency dependent magnetic permeability of sample A...D

CONCLUSION

Ba₃Co₂La$_x$Fe$_{24-x}$O$_{41}$ (x=0, 0.1, 0.5, 1.0) were made with the modified Pechini method followed by heat treatment at 600°C and calcination and sintering at 1300 and 1330°C respectively. Nearly pure Co₂Z (sample A) was obtained with uniform grain size distribution and a high permeability $\mu_A'$ = 11.6 at < 500MHz. A second phase Co₂Y occurs at x=0.1 (sample B) with larger grain size and lower permeability $\mu_B'$ = 9.7. With further increase of La doping concentration (x=0.5 and 1.0), the grain size decreases, in correlation with the decreasing permeability $\mu_C'$ = 9.1 and $\mu_D'$ = 8.0 for sample C and D respectively. The measured permeability spectra can be described by the superposition of domain wall motion and spin rotational magnetization. Both domain wall motion and spin rotational contributions are sensitive to the microstructures, average grain size and impurity phases of La-doped Co₂Z ferrites.

ACKNOWLEDGMENTS

This work was supported by Lockheed Martin Corporation under Grant Agreement/Contract No. CDA09-034-PE.

REFERENCES
1.  Bae, S.; Hong, Y. K.; Lee, J. J.; Park, J. H.; Jalli, J.; Abo, G. S.; Seong, W. M.; Park, S. H.; Kum, J. S.; Ahn, W. K.; Kim, G. H., Dual-Band Ferrite Chip Antenna for Global Positioning System and Bluetooth Applications. *Microw Opt Techn Let* **2011**, 53, (1), 14-17.
2.  Colom-Ustariz, J.; Rodriguez-Solis, R. A.; Velez, S.; Rodriguez-Acosta, S., Frequency agile microwave components using ferroelectric materials. *Sensors, Systems and Next-Generation Satellites Vi* **2003**, 4881, 280-286
3.  Smit, J., Wijn, H.P.J.,, *Ferrites*. John Wiley&Sons: Eindhoven, 1959.
4.  Kramar, G. P.; Panova, Y. I., Resonance and Relaxation of Domain-Walls in Polycrystalline Ferrites. *Phys Status Solidi A* **1983**, 77, (2), 483-489.
5.  Lubitz, P., New substitutions in hexagonal ferrites to reduce anisotropy without using Co. *J Appl Phys* **2000**, 87, (9), 4978-4980.
6.  Xu J., J. G., Zou H., Zhou Y., Gan S., Structural, Dielectric and Magnetic Properties of Nd-Doped Co2Z-Type Hexaferrites. *J Alloy Compd* **2011**, 509, 4290-4294.
7.  Gan, S. C.; Xu, J. J.; Yang, C. M.; Zou, H. F.; Song, Y. H.; Gao, G. M.; An, B. C., Electromagnetic and microwave absorbing properties of Co$_2$Z-type hexaferrites doped with La$^{3+}$. *J Magn Magn Mater* **2009**, 321, (19), 3231-3235.
8.  Guo S., Z. Y., Feng Z., Wang X., He H. , Magnetic Properties of Gd and Tb Doped Co$_2$Z Type Hexagonal Soft Magnetic Ferrites. *J Rare Earth* **2007**, 25, 220.
9.  Han, M. G.; Liang, D. F.; Deng, L. J., Analyses on the dispersion spectra of permeability and permittivity for NiZn spinel ferrites doped with SiO$_2$. *Appl Phys Lett* **2007**, 90, (19).
10.  Su, H.; Zhang, H. W.; Tang, X. L.; Jing, Y. L., Influence of microstructure on permeability dispersion and power loss of NiZn ferrite. *J Appl Phys* **2008**, 103, (9).
11.  Mu, C. L. Y. S., Y.; Wang, L.; Zhang, H., Improvement of high-frequency characteristics of Z-type hexaferrite by dysprosium doping. *J Appl Phys* **2011**, 109.

# ZINC OXIDE (ZnO) AND BANDGAP ENGINEERING FOR PHOTOELECTROCHEMICAL SPLITTING OF WATER TO PRODUCE HYDROGEN

Sudhakar Shet,[1, 2,] Yanfa Yan,[1] Heli Wang,[1] Nuggehalli Ravindra,[2] John Turner,[1] and Mowafak Al-Jassim[1]

[1] National Renewable Energy Laboratory, Golden, CO 80401 USA
[2] New Jersey Institute of Technology, Newark, NJ 07102 USA

ABSTRACT

We report material synthesis, characterization, and photoelectrochemical (PEC) measurements to explore methods to effectively reduce the band gap of ZnO for the application of PEC water splitting. We find that the band gap reduction of ZnO can be achieved by N and Cu incorporation into ZnO. We demonstrate that heavy Cu-incorporation lead to both p-type doping and band gap significantly reduced ZnO thin films. The p-type conductivity in our ZnO:Cu films is clearly revealed by Mott-Schottky plots. We further have successfully synthesized ZnO:N thin films with various reduced band gaps by reactive RF magnetron sputtering. The band gap reduction and photoresponse with visible light for N- and Cu-incorporated ZnO thin films are demonstrated.

INTRODUCTION

Transition metal oxide-based photoelectrochemical (PEC) splitting of water has attracted wide interest since photoinduced decomposition of water on $TiO_2$ electrodes was discovered.[1] To date, only $TiO_2$ has received extensive attention.[2] It is necessary to search for new metal oxides that can be potentially better light absorber. ZnO has similar bandgap and band-edge positions compared to $TiO_2$.[2] Furthermore, ZnO has a direct bandgap and higher electron mobility than $TiO_2$.[3] Thus, ZnO could also be a potential candidate for photoelectrochemical applications.[4] However, like $TiO_2$, the bandgap of ZnO (~3.3 eV) is too large to effectively use visible light. Hence, it is critical to reduce its bandgap to achieve a higher absorption coefficient. So far, impurity incorporation has been the main method to reduce the bandgap of $TiO_2$. Although band gap reduction of $TiO_2$ has been extensively studied, very limited research exists on bandgap narrowing of ZnO by impurity incorporation. Furthermore, ZnO is a native n-type semiconductor and is known to be difficult to make as p-type. For the application of water splitting, the use of both n-type and p-type semiconductors is often desirable.[5-12] Therefore, it is also important to synthesize both p-type and bandgap-reduced ZnO thin films. However, such investigations are lacking.

In this paper, we report on band gap reduction induced by incorporation of impurities, such as N and Cu. We have successfully synthesized heavy Cu-incorporation into ZnO lead to both p-type doping and band gap significantly reduced ZnO thin films. The p-type conductivity in our ZnO:Cu films is clearly revealed by Mott-Schottky plots. The bandgap reduction is demonstrated by UV-Vis absorption measurements. We further demonstrate that N –incorporation into ZnO (ZnO:N) thin films with various reduced band gaps by reactive RF magnetron sputtering.

MATERIALS SYNTHESIS

**ZnO:Cu thin films.** ZnO:Cu films were deposited using a reactive RF magnetron sputtering system followed by post-deposition annealing at 500 °C or 600 °C in air for 2 hrs. X-ray diffraction and AFM reveal that the grain size of the annealed pure ZnO film is much larger than that of the annealed ZnO:Cu films. Both the grain size and surface roughness of ZnO:Cu films decrease as the number of Cu chips increases. The Cu concentrations in ZnO:Cu films were measured by XPS, which increase from 3.0 to 9.8 at% with the increase of Cu chips from 4 to 12. After post-deposition annealing, the films exhibited a polycrystalline structure, with a main peak at 34.4° in the XRD patterns

231

corresponding to the (002) plane. There are no peaks corresponding to metallic Cu or its compounds, indicating no obvious phase separation in as-grown and 500 °C-annealed ZnO:Cu films. We found, however, that a CuO phase begins to form when ZnO:Cu films were annealed at 600°C. XPS data also indicate that there are no CuO peaks for the 500 °C-annealed ZnO:Cu films, indicating that Cu atoms exist in either the $Cu^0$ or $Cu^{1+}$ states. Both optical properties and PEC measurements indicate that Cu has the $Cu^{1+}$ states in the 500 °C annealed ZnO:Cu films and mixed $Cu^0$ and $Cu^{1+}$ states in the as-grown films. Figures 1(a) and 1(b) show the optical absorption spectra of as-grown and 500°C-annealed pure ZnO and ZnO:Cu films, respectively. It is seen that the absorption of the as-grown ZnO:Cu films extend into the infrared region. These absorptions may only be explained by the metallic Cu states. Because the depositions were carried out at room temperature, the ZnO:Cu films were not fully crystallized, as revealed by the XRD curves.

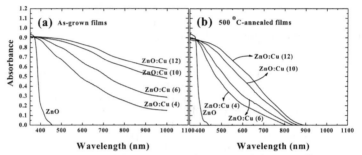

Fig. 1. Optical absorption spectra of (a) as-grown and (b) post annealed ZnO:Cu films.

It is very likely that Cu atoms may not be fully oxidized, leaving metallic Cu in the as-grown films. Such metallic Cu atoms are responsible for the absorption in the infrared region. However, the 500 °C-annealed ZnO:Cu films did not absorb the photons in the infrared region. The measured direct optical band gaps for ZnO:Cu films annealed 500°C gradually decreased from 3.16 to 3.05 eV with the increase of the Cu concentration. These reductions are attributed to up-shift of O $2p$ orbital, which is the result of the coupling between Cu $3d$ orbital and O $2p$ orbital.[13] In addition, significant absorption tails are also observed, as shown in Fig. 2, which can be attributed to the Cu generated impurity-bands.

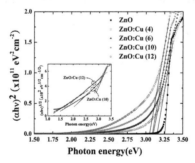

Fig. 2. Relative absorption coefficients of post annealed ZnO:Cu thin films.

Figure 3(a) shows the Mott-Schottky plot of the 500°C-annealed pure ZnO film. It has a positive slope in the linear region of the plot, indicating an n-type semiconductor. Figure 3(b) shows the Mott-Schottky plot of the 500°C-annealed ZnO:Cu(12) film, which had a negative slope indicating p-type behavior.

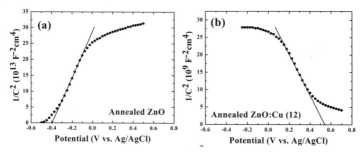

Fig. 3. Mott-Schottky plots of 500°C-annealed (a) pure ZnO and (b) ZnO:Cu(12) films.

It indicates that the $Cu^0$ metallic states were activated into the form of $Cu^{1+}$ substitutional acceptor states by the post-deposition annealing process at 500°C, because the $Cu^0$ metallic states cannot result in a p-type semiconductor. In a similar manner, all other 500°C-annealed ZnO:Cu films in this experiment showed negative slopes (not shown here). To confirm p-type conductivity, open-circuit voltage ($V_{oc}$), and PEC characteristics for the 500°C-annealed pure ZnO and ZnO:Cu films were also investigated (not shown here). The $V_{oc}$ value of the ZnO:Cu films increased with light illumination, whereas the $V_{oc}$ value of the pure ZnO decreased with light illumination. Moreover, the PEC characteristics of the ZnO:Cu films under chopped light illumination showed cathodic photoresponses (direct indicative of p-type semiconductor), whereas the pure ZnO film exhibited anodic photoresponses (n-type semiconductor). These investigations confirm that the 500°C-annealed ZnO:Cu films are p-type semiconductors.

**ZnO:N films.** To synthesize ZnO:N films with similar thickness, we first determined the deposition rate for different RF powers. We found that the deposition rate increases linearly with the increase of RF power. Further, the gas ambient has a great impact on the deposition rate. We found that the deposition rate of ZnO:N films grown in mixed $N_2$ and $O_2$ ambient is much higher than that of ZnO films grown in pure $O_2$ ambient, even though the same RF power was used. On the other hand, the deposition rate of the ZnO:N films grown in mixed $N_2$ and $O_2$ ambient at 200 W is similar to that of the $Zn_3N_2$ films grown at 200 W. The enhancement of the deposition rate in $N_2$-containing ambient could be due to the nitridation of the Zn metal target surface, because the Zn nitride is more conductive than the ZnO, leading to much higher sputtering yield. XRD revealed that the ZnO film exhibits poor crystallinity, due to the room-temperature sputtering process. The ZnO:N film grown at 80 W shows better crystallinity than the pure ZnO film, despite the faster deposition rate. It may be due to the significant change in the chamber ambient. For pure ZnO growth, the ambient is pure $O_2$ gas. For ZnO:N growth, the ambient is mainly $N_2$ with only 5% $O_2$. When the RF power was increased to 100 W, the crystallinity again became poor. When the RF power was further increased to 200 W, the film became $Zn_3N_2$-like. The sample grown at 200 W became polycrystalline $Zn_3N_2$-like.

The microstructure of the ZnO:N films was investigated by TEM. We found that the ZnO:N films have an intermediate ordering structure. The grain sizes are very small, in the range of a few nanometers. Nano-probe X-ray energy dispersive (EDS) revealed that these small grains are indeed ZnO:N with high N concentration. Quantitative EDS analysis revealed that the N concentration is about 2, 18, 29, 35 at% for samples at RF powers of 80, 100, 120, and 150 W, respectively. However, it gives an indication that the N concentration increases with the increase of RF powers.

Figure 4(a) shows the relative absorption coefficients for the ZnO:N films grown at different RF powers. The absorption coefficient was assumed to evaluate the band gap energies of the films. Further, direct electron transition from the valence to the conduction bands was also assumed for the absorption coefficient curves, since both ZnO and $Zn_3N_2$ are known as direct-band gap materials. Figure 4(b) shows the absorption spectra of ZnO:N films grown at different RF powers. The absorption spectra show that the light absorptions are successfully shifted into visible regions with the increase of RF power. The measured band gaps are 2.9 eV (RF power = 80 W), 2.31 eV (RF power = 100 W), 2.15 eV (RF power = 120 W), 1.76 eV (RF power = 150 W), and 1.55 eV (RF power = 200 W).

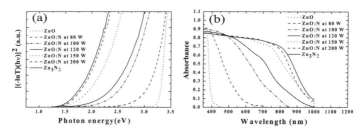

Fig. 4. (a) Relative absorption coefficients of pure ZnO, $Zn_3N_2$, and ZnO:N films grown at different RF powers. (b) Optical absorption spectra of the ZnO:N films grown at different RF powers.

The band gaps are 3.28 eV and 1.52 eV for pure ZnO and $Zn_3N_2$, respectively. Therefore, we demonstrate that RF power could be used effectively to control the N concentration in ZnO thin films. We also found that $Zn_3N_2$ film is not stable. It can be oxidized in air easily.

PEC properties were measured on ZnO:N films grown at a RF power of 80 W and as-grown and annealed pure ZnO thin films. The positive slopes in the Mott-Schottky plots of these thin films indicate that all of the samples are n-type semiconductors.

It is well known that ZnO films are native n-type. The origins of the n-type conductivity in the as-grown and annealed ZnO films are likely due to native defects such as Zn interstitials or/and oxygen vacancies $(V_o)$.[14] It is known that an N occupying at an O site $(N_o)$ is an acceptor, which should make ZnO:N p-type. However,. Our ZnO:N films were sputtered under the $N_2/O_2$ plasma. It has been reported that the ZnO films deposited in $N_2$ plasma were usually n-type, due to substitutional $N_2$ molecules at an O site that act as shallow double donors.

Figure 5 shows the photocurrent-voltage curves of the samples under illumination with UV/IR and (a) blue filter, (b) yellow filter, and (c) red filter, respectively. The insets show the filtered optical spectra.

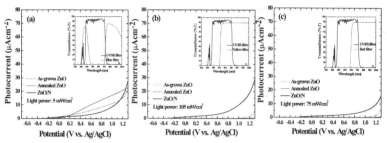

Fig. 5. Photocurrent-voltage curves of the samples under light illumination with the UV/IR filter and color filters: (a) blue, (b) yellow, (c) red filters. The inserts show the profiles of the filter.

The photocurrents of the as-grown and annealed ZnO films in Fig. 5(a) are almost the same as that without UV/IR filter. Since ZnO films have large band gaps and can only absorb light in the short-wavelength region below 450 nm, the blue filter should pass all above gap photons leaving photocurrents unaltered. However, with blue filtering, the photocurrent of the ZnO:N film is smaller than that of the ZnO films at potentials below 1.1V. In the long-wavelength regions, the ZnO:N film exhibits photoelectrochemical property, but the ZnO films do not due to its large band gap as it would be expected for a lower band gap material. Figs. 5(b) and (c) show that both the as-grown and annealed ZnO films do not demonstrate any photoresponse in the long-wavelength region above 500 nm, because pure ZnO films do not absorb any light in long-wavelength region above 450 nm. However, the ZnO:N film showed clear photoresponse in the long-wavelength regions, even in the regions beyond 600 nm, as seen in Fig. 5(c). It should be noted that our ZnO:N films are not optimized. The photocurrent generated by our ZnO:N films are not very high. We also encounter instability issue with our films. Nonetheless, our results demonstrate that N-incorporation in ZnO can shift photoresponse of ZnO films into the visible light region, the main component of the sun light.

CONCLUSIONS

We have reported investigation on band gap reduction of ZnO by N and Cu incorporation into ZnO. We have successfully synthesized ZnO:N thin films with various reduced band gaps by reactive RF magnetron sputtering. PEC response in long wave length regions in ZnO:N films has been demonstrated. We have also shown that heavy Cu-incorporation lead to both p-type doping and significantly band gap reduced ZnO thin films. The band gap reduction is demonstrated by UV-Vis absorption measurements. The p-type conductivity in ZnO:Cu films have been demonstrated by Mott-Schottky plots.

REFERENCES

[1] K. Honda and A. Fujishima, *Nature* 238, p. 35, 1972.

[2] M. Grätzel, *Nature* 414, p. 338, 2001.

[3] K. Kakiuchi, E. Hosono, and S. Fujihara, *J. Photochem. & Photobiol. A: Chem.* 179, p. 81, 2006.

[4] T.F. Jaramillo, S.H. Baeck, A. Kleiman-Shwarsctein, and E.W. McFarland, *Macromol. Rapid Comm.* 25, p. 297, 2004.

[5] S. Shet, K. –S. Ahn, Y. Yan, T. Deutsch, K. M. Chrusrowski, J. Turner, M. Al-Jassim, and N. Ravindra, *J. Appl. Phys.* 103, p. 073504, 2008.

[6] K. –S. Ahn, S. Shet, T. Deutsch, C. S. Jiang, Y. Yan, M. Al-Jassim, and J. Turner,

*J. Power Source*, 176, p. 387, 2008.

[7] S. Shet, K. –S. Ahn, T. Deutsch, H. Wang, N. Ravindra, Y. Yan, J. Turner, M. Al-Jassim, *J. Mater. Research* 25, 69 Doi: 10.1557/JMR.2010.0017, 2010.

[8] S. Shet, K.-S. Ahn, T. Deutsch, H. Wang, N. Ravindra, Y. Yan, J. Turner, M. Al-Jassim, *J. Power Sources* 195, p. 5801, 2010.

[9] K.–S. Ahn, Y. Yan, S. Shet, T. Deutsch, J. Turner, and M. Al-Jassim, *Appl. Phys. Lett.* 91, p. 231909, 2007.

[10] S. Shet, K. –S. Ahn, H. Wang, N. Ravindra, Y. Yan, J. Turner, M. Al-Jassim, *J. Mater. Science* DOI 10.1007/s10853-010-4561-x, 2010.

[11] K.-S. Ahn, Y. Yan, S. Shet, K. Jones, T. Deutsch, J. Turner, M. Al-Jassim, *Appl. Phys. Lett.* 93, p. 163117, 2008.

[12] S. Shet, K. –S. Ahn, N. Ravindra, Y. Yan, J. Turner, M. Al-Jassim, *J. Materials* 62, p. 25, 2010.

[13] Y. Yan, M. M. Al-Jassim, and S. H. Wei, *Appl. Phys. Lett.* 89, p. 181912, 2006.

[14] K. K. Kim, H. S. Kim, D. K. Hwang, J. H. Lim, and S. J. Park, *Appl. Phys. Lett.,* 83, p. 63, 2003.

# INVESTIGATION OF ZnO:N AND ZnO:(Al,N) FILMS FOR SOLAR DRIVEN HYDROGEN PRODUCTION

Sudhakar Shet,[1, 2,] Yanfa Yan,[1] Nuggehalli Ravindra,[2] Heli Wang,[1] John Turner,[1] and Mowafak Al-Jassim[1]

[1] National Renewable Energy Laboratory, Golden, CO 80401 USA
[2] New Jersey Institute of Technology, Newark, NJ 07102 USA

ABSTRACT

ZnO thin films with significantly reduced bandgaps were synthesized by doping N and co-doping Al and N at 100°C. All the films were synthesized by radio-frequency magnetron sputtering on F-doped tin-oxide-coated glass. We found that co-doped ZnO:(Al,N) thin films exhibited significantly enhanced crystallinity as compared to ZnO doped solely with N, ZnO:N, at the same growth conditions. Furthermore, annealed ZnO:(Al,N) thin films exhibited enhanced N incorporation over ZnO:N films. As a result, ZnO:(Al,N) films exhibited improved photocurrents than ZnO:N films grown with pure N doping.

INTRODUCTION

Transition metal oxide-based photoelectrochemical (PEC) splitting of water has attracted wide interest since photoinduced decomposition of water on $TiO_2$ electrodes was discovered.[1] To date, most investigations have focused on $TiO_2$.[2] The drawback of a PEC system using $TiO_2$ is that it can only absorb ultraviolet (UV) light due to its large bandgap of 3.0–3.2 eV. Therefore, it is necessary to search for new metal oxides that can potentially absorb visible light. ZnO has similar bandgap and band-edge positions compared to $TiO_2$.[2] Furthermore, ZnO has a direct bandgap and higher electron mobility than $TiO_2$.[3] Thus, ZnO could also be a potential candidate for photoelectrochemical applications.[4] However, like $TiO_2$, the bandgap of ZnO (~3.3 eV) is too large to effectively use visible light. Hence, it is critical to reduce its bandgap to achieve a higher absorption coefficient.

It has been suggested that doping with C, S, and N would result in a reduced bandgap for most wide-bandgap metal oxides. Results for experimental incorporation of N into $TiO_2$, ZnO, and $WO_3$ have been reported.[4-5] Significant amounts of N can only be incorporated into ZnO and $WO_3$ at low temperatures.[5] However, the films grown at low temperatures usually exhibit a very poor crystallinity, which is extremely detrimental to the PEC performance. This dilemma hinders the PEC performance of N-incorporated ZnO and $WO_3$ films. A possible cause for the inferior crystallinity is due to the uncompensated charged N atoms. This problem could be overcome by charge-compensated donor-acceptor doping, such as, co-doping ZnO with Al and N. Furthermore, incorporating (Al,N) pairs is easier than incorporating sole N atoms because of donor-acceptor interaction.[6-9] Most of the studies on Al, N co-doped ZnO thin film focused on the $p$-type doping issue. The effect of co-doping bandgap reduction and PEC performance has not been investigated. In this paper, we report on the synthesis, crystallinity, bandgap reduction, and PEC response of Al, N co-doped ZnO, (ZnO:Al:N) or ZnO: (Al, N) thin films compared to ZnO, Al doped ZnO, ZnO:Al, and N doped ZnO, ZnO:N films..

EXPERIMENTAL

The ZnO:Al and ZnO:(Al,N) thin films were deposited by reactive rf sputtering of a ZnO-2wt%Al target using an oxygen/nitrogen gas mixture. Hereafter, we will refer ZnO:(Al,N) thin films as ZnO:Al;N in the text and figures. For ZnO and ZnO:N thin films, ZnO target was used. All ZnO thin films were grown on F-doped tin oxide (FTO) (20–23 $\Omega/\square$)-coated glass to allow PEC measurements. Substrates were rotated at 30 RPM during deposition to enhance film uniformity. The base pressure was below $5\times10^{-6}$ torr and working pressure was $2\times10^{-2}$ torr. The sputtering ambient was mixed $O_2$ and $N_2$ with an oxygen gas ratio $O_2/(N_2+O_2) = 2.5\%$. After pre-sputtering for 30 min, sputtering was conducted at a RF power of 300 W. All the samples were grown at $100^{\circ}C$ and annealed at $500^{\circ}C$ in air for 2 h. All samples were controlled to have a similar film thickness of about 1000 nm as measured by stylus profilometry.

The UV-Vis absorption spectra of the samples were measured to investigate the optical properties. PEC measurements were performed in a three-electrode cell with a flat quartz window.[7-18] The ZnO films (area: $0.25$ cm$^2$) were used as the working electrodes. A Pt sheet (area: 10 cm$^2$) and a Ag/AgCl electrode were used as counter and reference electrodes, respectively. A 0.5-M Na$_2$SO$_4$ aqueous solution was used as the electrolyte.[19-24] PEC response was measured using a fiber-optic illuminator (tungsten-halogen lamp) with a UV/IR cut-off filter (cut-off wavelengths: 350 and 750 nm) and combined UV/IR and green band-pass filter (wavelength: 538.33 nm; full width at half maximum: 77.478 nm). Total light intensity with the UV/IR filter was fixed at 125 mW/cm$^2$.[10-22]

The electrical properties were measured using Mott-Schottky plots obtained by AC impedance measurements.[25] AC amplitude of 10 mV and an AC frequency of 5000 Hz were used under dark conditions and the AC impedances were measured in the potential range of -0.7 to 1.4 V. A simple equivalent-circuit model of resistance and capacitance in series was used to analyze the Mott-Schottky plots.[25]

RESULTS AND DISCUSSION

Figure 1 shows the XRD curves for ZnO, ZnO:Al, ZnO:N, and ZnO:Al;N films. The ZnO, ZnO:Al, and ZnO:N films exhibited poor crystallinity. The crystalline quality was characterized from the full width at half maximum (FWHM) of the XRD curve instead of the peak intensity. The crystallite size, estimated according to the Scherrer equation using the FWHM of (002) peak. The crystallite size was estimated to be about 16 nm, 18nm, and 22 nm for ZnO, ZnO:Al, and ZnO:N respectively. X-ray photoelectron spectroscopy (XPS) results (not shown here) confirmed that no significant amount of nitrogen (at.%) is present in the ZnO:N films. These results indicate that it is very difficult to incorporate a significant amount of N into ZnO by sputtering a ZnO target. Also these films have poor crystallinity because of deposition at low substrate temperatures. ZnO films with high crystallinity can be synthesized at high temperatures; but at these substrate temperatures, it is more difficult to incorporate a significant amount of nitrogen into the films. Surprisingly, however, the co-doped ZnO:Al:N films exhibited significantly enhanced crystallinity, yet with the significant amount of incorporation of N.

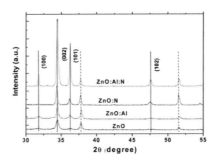

Fig. 1. XRD curves for a ZnO, ZnO:Al, ZnO:N, and ZnO:Al;N films, respectively.

From the XRD curve shown in Fig. 1, the crystallite size of ZnO:Al:N films is estimated to be around 44 nm. Such significantly enhanced crystallinity could be attributed to the co-doping effect and additional oxygen supplied in the chamber ambient during the sputtering process, leading to decreased oxygen vacancies and enhanced crystallinity. The nitrogen concentration in the ZnO:Al:N films was estimated to be about (5 at.%), which is much higher than that of the ZnO:N film. This is because the charge compensated donor-acceptor co-doping, which reduced the formation energy helped in incorporating more N concentration.

Figure 2(a) shows optical absorption spectra of the annealed ZnO, ZnO:Al, ZnO:N, and ZnO:Al:N thin films. The ZnO, and ZnO:Al films showed similar optical absorption spectra and could absorb only light with wavelength below 450 nm, due to their wide bandgap. However, the ZnO:N film could absorb lower energy photons, up to 1000 nm, indicating that the more oxygen vacancies and bandgap was narrowed by N incorporation in ZnO. However, ZnO:Al:N film exhibited best absorption curve, indicating that the bandgap was significantly reduced by much higher concentration of N incorporated in ZnO. Figure 2(b) shows the optical absorption coefficients for the samples used in this experiment. The direct electron transition from valence to conduction bands was assumed to calculate the bandgap from the absorption coefficient curves, because ZnO films are known to be direct-bandgap materials. The optical bandgaps of the films were determined by extrapolating the linear portion of each curve. The bandgap of the ZnO and ZnO:Al films are 3.26 and 3.35 eV respectively, which corresponds well to the bandgap of the ZnO and Al doped ZnO (ZnO:Al). The ZnO:N film exhibited a decreased bandgap of 3.19 eV, due to the N incorporation. The co-doped ZnO:Al:N film exhibited a significantly reduced bandgap of 2.02 eV, indicating high concentration of N is incorporated in the film.

This large reduction in bandgap is due to N-induced upshifting of the valance band. It is shown theoretically that the incorporated N would generate an impurity band above the valance band. The absorption from this impurity band cannot be characterized by direct band transitions and typically results in an absorption tail in the measured optical absorption curve. Such an absorption tail is clearly evident in Fig. 2 (b) for the co-doped ZnO:Al:N film. This tail can be considered further bandgap reduction, which enables light harvesting in the much longer wavelength regions compared to the ZnO, ZnO:Al, and ZnO:N films.

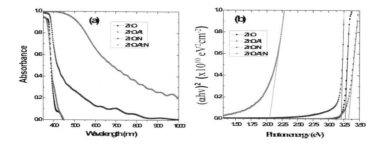

Fig.2. (a) Optical absorption curves ZnO, ZnO:Al, ZnO:N, and ZnO:Al;N films respectively. (b) Direct transition-optical absorption coefficients of a ZnO, ZnO:Al, ZnO:N, and ZnO:Al;N films respectively.

Figure 3 shows Mott-Schottky plots of ZnO, ZnO:Al, ZnO:N, and Zno:Al:N films. All the films exhibited positive slopes, indicating $n$-type semiconductors. Our previous studies [7-9, 11-14] reported that ZnO:N films deposited under a $N_2/O_2$ plasma showed $n$-type behavior, due to substitutional $N_2$ molecules that act as shallow double-donors.

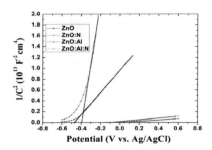

Fig. 3. Mott-Schottky plots of ZnO, ZnO:Al, ZnO:N, and Zno:Al:N films.

Figure 4(a) and (b) shows the photocurrent-voltage curves of ZnO and ZnO:Al:N films under light on/off illumination with the UV/IR filter. It showed clearly that the ZnO:Al:N film exhibited significantly increased photocurrents, compared to the ZnO and other films(not shown here).

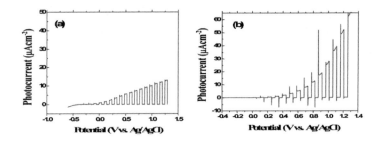

Fig. 4. (a) Photocurrent-voltage curves of a ZnO film under light on/off illumination with the UV/IR filter. (b) Photocurrent-voltage curves of a ZnO:Al:N film under light on/off illumination with the UV/IR filter.

Figure 5(a) shows the photocurrent-voltage curves of ZnO, ZnO:Al, ZnO:N, and ZnO:Al:N films under illumination with the UV/IR filter. It showed clearly that the ZnO:Al:N film exhibited enhanced photocurrents, compared to the other films. At the potential of 1.2 V, the photocurrents were 11.6, 14.6, 15, and 54 $\mu Acm^{-2}$ for the ZnO, ZnO:Al, ZnO:N, and Zno:Al:N films, respectively. To investigate the photoresponses in the long-wavelength region, a green color filter (wavelength: 538.33 nm; FWHM: 77.478 nm) was used in combination with the UV/IR filter, as shown in Fig. 5(b). The ZnO and ZnO:Al films exhibited no clear photoresponse, due to its wide bandgap. The co-doped ZnO:Al:N film exhibited much higher photocurrent than the ZnO:N film, despite much less light absorption. It indicates that a very high recombination rate of the photogenerated electrons and holes is present in the ZnO:N film, due to its inferior crystallinity, and uncompensated charges.

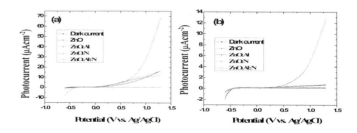

Fig. 5. Photocurrent-voltage curves of a ZnO, ZnO:Al, ZnO:N, and ZnO:Al;N films respectively, under the illumination (a) with an UV/IR filter and (b) with the combined green and UV/IR filters.

On the other hand, the co-doped ZnO:Al:N film exhibited remarkably increased crystallinity, and charge compensation, which lead to enhanced photocurrent than the other films. The results demonstrate clearly that significantly reduced bandgap and enhanced photocurrents can be obtained from charge compensated co-doping approach. We expect that further enhanced photocurrents should be possible with greater N incorporation in ZnO and better charge compensation facilitated by the co-doping approach.

CONCLUSIONS

We report on the bandgap narrowing and photoelectrochemical (PEC) response of Al, N co-doped ZnO thin films. The ZnO.Al:N thin films were deposited by sputtering at substrate temperature of 100°C and followed by postannealing at 500°C in air for 2 hours. We found that ZnO:Al:N thin films exhibited significantly enhanced crystallinity compared to ZnO, Al doped ZnO (ZnO:Al), and N doped ZnO (ZnO:N) at the same growth conditions. Furthermore, ZnO:Al:N thin films exhibited enhanced N-incorporation resulted in much reduced bandgap. As a result, ZnO:Al:N thin films achieved improved PEC response, compared to ZnO, ZnO:Al, and ZnO:N thin films.

REFERENCES

[1] K. Honda and A. Fujishima, *Nature (London)* 238, p. 37, 1972.

[2] M. Grätzel, *Nature* 414, p. 338, 2001.

[3] K. Kakiuchi, E. Hosono, and S. Fujihara, *J. Photochem. & Photobiol. A: Chem.* 179, p. 81, 2006.

[4] T.F. Jaramillo, S.H. Baeck, A. Kleiman-Shwarsctein, and E.W. McFarland, *Macromol. Rapid Comm.* 25, p. 297, 2004.

[5] D. Paluselli, B. Marsen, E. L. Miller, and R. E. Rocheleau, *Electrochem. and Solid-State Lett.* 8, p. G301, 2005.

[6] Y. Yan, S. B. Zhang, and S. T. Pantelides, *Phys. Rev. Lett.* 86, p. 5723, 2001.

[7] S. Shet, K. –S. Ahn, T. Deutsch, H. Wang, N. Ravindra, Y. Yan, J. Turner, M. Al-Jassim, *J. Mater. Research* 25, 69 Doi: 10.1557/JMR.2010.0017, 2010.

[8] K.–S. Ahn, Y. Yan, S. Shet, T. Deutsch, J. Turner, and M. Al-Jassim, *Appl. Phys. Lett.* 91, p. 231909, 2007.

[9] S. Shet, K. –S. Ahn, H. Wang, N. Ravindra, Y. Yan, J. Turner, M. Al-Jassim, *J. Mater. Science* DOI 10.1007/s10853-010-4561-x, 2010.

[10] S. Shet, K. –S. Ahn, Y. Yan, T. Deutsch, K. M. Chrusrowski, J. Turner, M. Al-Jassim, and N. Ravindra, *J. Appl. Phys.* 103, p. 073504, 2008.

[11] K. –S. Ahn, S. Shet, T. Deutsch, C. S. Jiang, Y. Yan, M. Al-Jassim, and J. Turner, *J. Power Source*, 176, p. 387, 2008.

[12] S. Shet, K.-S. Ahn, T. Deutsch, H. Wang, N. Ravindra, Y. Yan, J. Turner, M. Al-Jassim, *J. Power Sources* 195, p. 5801, 2010.

[13] K.-S. Ahn, Y. Yan, S. Shet, K. Jones, T. Deutsch, J. Turner, M. Al-Jassim, *Appl. Phys. Lett.* 93, p. 163117, 2008.

[14] S. Shet, K. –S. Ahn, N. Ravindra, Y. Yan, J. Turner, M. Al-Jassim, *J. Materials* 62, p. 25, 2010.

[15] H. Wang, T. Deutsch, S. Shet, K. Ahn, Y. Yan, M. Al-Jassim and J. Turner, *Solar Hydrogen and Nanotechnology IV, SPIE*, Nanoscience + Engineering, p. 7408, 2009.

[16] S. Shet , K. Ahn, N. Ravindra, Y. Yan, T. Deutsch, J. Turner, M. Al-Jassim *Proceedings of the Materials Science & Technology*, p. 219, 2009.

[17] S. Shet, K. Ahn, N. Ravindra, Y. Yan, T. Deutsch, J. Turner, M. Al-Jassim *Proceedings of the Materials Science & Technology*, p. 277, 2009.

[18] K.-S. Ahn, Y. Yan, M.-S. Kang, J.-Y. Kim, S. Shet, H. Wang, J. Turner, and M. Al-Jassim, *Appl. Phys. Lett.* 95 p. 022116, 2009.

[19] Y. Yan, K. Ahn, S. Shet, T. Deutsch, M. Huda, S. Wei, J. Turner, M. Al-Jassim, *Proceedings of the SPIE*, 6650, p. 66500H, 2007.

[20] S. Shet, K. Ahn, N. Ravindra, Y. Yan, T. Deutsch, J. Turner, M. Al-Jassim, Materials Science & Technology 2009, *Ceramic Transactions volume*, (2010) in press

[21] K.-S. Ahn, S. Shet, Y. Yan, J. Turner, M. Al-Jassim, N. M. Ravindra, *Proceedings of the Materials Science & Technology*, p. 901, 2008.

[22] S. Shet, K. –S. Ahn, N. Ravindra, Y. Yan, J. Turner, M. Al-Jassim, *J. Materials* 62, p. 25, 2010.

[23] K. Ahn, S. Shet, T. Deutsch, Y. Yan, J. Turner, M. Al-Jassim, N. M. Ravindra, *Proceedings of the Materials Science & Technology*, p. 952, 2008.

[24] S. Shet, K. Ahn, T. Deutsch, Y. Yan, J. Turner, M. Al-Jassim, N. Ravindra, *Proceedings of the Materials Science & Technology,* p. 920, 2008.

[25] J. Akikusa and S. U. M. Khan, *Int. J. Hydrogen Energy* 27, p. 863, 2002.

# Author Index

Al-Jassim, M., 231, 237

Ballato, J., 113
Basu, B., 191
Berghmans, A., 65
Bhalla, A., 179, 191
Bhalla, A. S., 55
Bledt, C. M., 3
Bobnar, V., 23

Cao, Y., 125
Chae, K.-W., 167
Cheon, C.-I., 167
Childers, W., 147
Chumanov, G., 113
Colorado, H. A., 13
Cótica, L. F., 209, 215

Das, D., 77
Dias, G. S., 209, 215
Dubey, A. K., 191
Dukic, M., 199

Eiras, J. A., 209, 215
Eršte, A., 23
Estrada, F. R., 215

Freitas, V. F., 215

Garcia, D., 209, 215
Gardner, P., 147

Gotardo, R. A. M., 209
Govindaraju, N., 77, 87
Green, M. L., 31
Guo, J. H., 125
Guo, R., 43, 55, 179, 191

Harrington, J. A., 3
He, Z. X., 125
House, M., 65
Huang, Q., 31

Jakub Cajzl, J., 95
Joress, H., 31
Jung, J., 43

Kagomiya, I., 167
Kahler, D., 65
Kim, J.-S., 167
King, M., 65
Knuteson, D., 65
Kolitsch, A., 95
Kopp, D. V., 3
Kosec, B., 23
Kosel, P. B., 77, 87
Kužnik, B., 23

Lee, Y.-J., 167
Li, J., 179
Li, Y., 179
Liu, J., 135
Lowhorn, N., 31